# ADAPTIVE THERMAL COMFORT

The fundamental function of buildings is to provide safe and healthy shelter. For the fortunate they also provide comfort and delight. In the twentieth century comfort became a 'product' produced by machines and run on cheap energy. In a world where fossil fuels are becoming ever scarcer and more expensive, and the climate more extreme, the challenge of designing comfortable buildings today requires a new approach.

This timely book is the first in a trilogy from leaders in the field which will provide just that. It explains, in a clear and comprehensible manner, how we stay comfortable by using our bodies, minds, buildings and their systems to adapt to indoor and outdoor conditions, which change with the weather and the climate. The book is in two sections. The first introduces the principles on which the theory of adaptive thermal comfort is based. The second explains how to use field studies to measure thermal comfort in practice and to analyse the data gathered.

Architects have gradually passed responsibility for building performance to service engineers who are largely trained to see comfort as the 'product', designed using simplistic comfort models. The result has contributed to a shift to buildings that use ever more energy. A growing international consensus now calls for low-energy buildings. This means designers must first produce robust, passive structures that provide occupants with many opportunities to make changes to suit their environmental needs. Ventilation using free, natural energy should be preferred and mechanical conditioning only used when the climate demands it.

This book outlines the theory of adaptive thermal comfort that is essential to understand and inform such building designs. This book should be required reading for all students, teachers and practitioners of architecture, building engineering and management – for all who have a role in producing, and occupying, twenty-first-century adaptive, low-carbon, comfortable buildings.

**Fergus Nicol** has led a number of important research projects on comfort, which have influenced thinking internationally. He has authored numerous journal articles and other publications including guidance on comfort and overheating. Nicol convenes the Network for Comfort and Energy use in Buildings and organises their regular international Windsor Conferences.

**Michael Humphreys** is known for his pioneering work on the adaptive approach to comfort. He has been Head of Human Factors at the Building Research Establishment and a Research Professor at Oxford Brookes University, UK. His current interests are the structure and modelling of human adaptive behaviour, the interactions between aspects of the environment and their expression in standards.

**Susan Roaf** wrote her PhD on comfort and the windcatchers of Yazd and after a decade working with Nicol and Humphreys at the Oxford Thermal Comfort Unit she is now Professor of Architectural Engineering at Heriot-Watt University in Edinburgh. She is a teacher, researcher, designer and author and editor of 13 books including *Ecohouse: A Design Guide* (2001) and *Adapting Buildings and Cities for Climate Change* (2005).

# ADAPTIVE THERMAL COMFORT

## Principles and practice

*Fergus Nicol, Michael Humphreys and Susan Roaf*

Routledge
Taylor & Francis Group

LONDON AND NEW YORK

First published 2012
by Routledge
2 Park Square, Milton Park, Abingdon, Oxon OX14 4RN

Simultaneously published in the USA and Canada
by Routledge
711 Third Avenue, New York, NY 10017

*Routledge is an imprint of the Taylor & Francis Group, an informa business*

*British Library Cataloguing in Publication Data*
A catalogue record for this book is available from the British Library

*Library of Congress Cataloging in Publication Data*
Nicol, F. (Fergus)
Adaptive thermal comfort : principles and practice / Fergus Nicol, Michael A.
Humphreys, and Susan Roaf.
p. cm.
Includes bibliographical references and index.
1. Buildings--Environmental engineering. 2. Buildings--Thermal properties. 3.
Architecture--Human factors. 4. Architecture and climate. I. Humphreys, Michael A.
(Michael Alexander), 1936- II. Roaf, Susan. III. Title.
TH6025.N53 2012
697--dc23

ISBN: 978-0-415-69159-8 (pbk)
ISBN: 978-0-203-12301-0 (ebk)

Typeset in Bembo
by GreenGate Publishing Services, Tonbridge, Kent
Printed and bound by CPI Group (UK) Ltd, Croydon CR0 4YY

For Gill

# CONTENTS

# ILLUSTRATIONS

## Figures

## Colour plate section (between pages 74 and 75)

## Tables

# PREFACE

One of the most important attributes of a building, after ensuring that it stays standing, is that it provides a thermally safe haven for its occupants in the climate and environment where it is built. The primary function of a building is to provide shelter. As buildings evolved over millennia this requirement for shelter merged into the desire for comfort and, in some climates and societies, the aspiration to live in the luxury of thermal delight. Think of a Mediterranean villa with its vine-covered terrace facing the cooling sea breeze, or the exquisitely decorated wind pavilion on a Persian desert palace on a summer's evening. Think of a warm fireside on a cold evening or a sun-filled winter garden on a snowy day. The art and science of great design lies in being able to generate such profoundly enjoyable experiences, first created millenia ago, and to work with the minimum use of fossil fuels. Great design requires a genuine understanding of comfort and how to achieve it.

This book has been written to educate readers in the theory of adaptive thermal comfort, methodologies for use in research and the tools, strategies and understanding needed to interpret and apply research findings. The audience we hope for includes students, particularly at Master's and PhD levels, building design professionals and researchers. The latter group increasingly includes those studying the phenomena related to climate change and how we adapt to avoid its worst impacts on buildings and cities. The overheating of buildings is recognised as one of the most lethal of the emerging properties of climate change as buildings increasingly fail to provide adequate shelter in extreme temperatures. Architects, in particular, and building services and architectural engineers should not begin their design careers without a firm grasp of this subject. It is essential to our ability to design increasingly resilient buildings.

A good understanding of adaptive thermal comfort is also necessary for anyone involved in the design of low-carbon buildings. Designing to reduce the need to heat and cool a building requires a greater understanding of how to achieve thermal comfort in buildings without using energy, or using it in such a way that its cooling or heating potential is maximised. As energy prices soar, owing to decreasing global supplies of cheap fossil fuel energy, the need to reduce energy use becomes a fundamental driver in commercial and domestic building markets. Buildings that are energy profligate are already being sidelined because of their

unaffordable running costs. Buildings can be run on clean, renewable energy, but only if they are low-energy buildings to begin with, and to be saleable they must be comfortable too.

So why this book now? There is no other good introductory handbook on the subject of adaptive thermal comfort. Most detailed specialist books do not cover the adaptive comfort approach adequately. Why is the adaptive approach so important? Because it is the only method of calculating and designing for desired indoor temperatures with occupant controlled naturally ventilated buildings for much of the year; because this significantly reduces energy use through tracking outdoor temperatures; and because it enables building occupants to take more control of their own comfort conditions.

There has been a growing dissatisfaction with climate-chamber-based heat-balance models for predicting the conditions that people will find thermally comfortable in buildings. These models are rather blunt tools. For instance, they require a knowledge of clothing and activity that building designers cannot guarantee and give little guidance on how to quantify for future situations. Thus designers are forced to make assumptions about people's future behaviour based on very little information, driving designers to favour simplistic, easy solutions such as choosing static design indoor temperatures. How many building operators today do not even change their indoor set-point temperature between winter and summer? The consequence is often high levels of energy use to over-cool buildings in summer and overheat them in winter, all too often resulting in uncomfortable if not unhealthy buildings (Mendell and Mirer, 2009).

Indoor temperature control currently uses around one quarter of all energy used in developed economies – and the proportion is growing. The increasing reliance on air conditioning has also led to an appreciable decline in the standard of passive, environmental design skills among architects. Such concerns mean that many people are increasingly questioning existing temperature standards, the methods by which they were derived and the processes by which they are applied.

Is it right for a generation of architects to know nothing about how to design buildings to be comfortable? The architectural profession has increasingly shifted the responsibility for the thermal performance of buildings to the engineering profession who are often only trained to be expert in mechanical solutions. A new generation of designers of low-carbon buildings is emerging with a broader skills-set, encompassing both architecture and engineering training. The old comfort standards do not work well for the new generation of buildings they are producing, but adaptive thermal comfort standards liberate the designer to create buildings that use natural energy from wind and sun when appropriate, to run buildings for much of the year on renewable energy and to open the windows again and restore the thermal delight that is too often absent in modern tight-skinned buildings.

Is this trilogy on adaptive comfort too late? The ideas behind the books trace their origin back to a collaboration between Michael Humphreys and Fergus Nicol in the 1960s and 1970s. Charles Webb had set up a field study on the comfort of office workers at the Building Research Establishment (BRE) in 1965, a project that Humphreys and Nicol inherited on his retirement. In analysing the results they became increasingly unhappy with the approaches at hand for the analysis of such data and developed their own approach to the subject, first fully presented at the first international symposium (Langdon et al., 1973) on thermal comfort at the Building Research Establishment (BRE) in 1972 (Nicol and Humphreys, 1973). In subsequent papers, until he left the BRE in 1978, Humphreys went on to develop and confirm the approach which has subsequently become known as the adaptive approach to thermal comfort.

In the 1980s Andres Auliciems and Richard de Dear in Australia (Auliciems, 1986; Auliciems and de Dear, 1990) carried the ideas forward. In the early 1990s Nicol, Humphreys and Roaf formed the Thermal Comfort Unit at Oxford Brookes University, and were part of a great deal of subsequent research, debate and regulation undertaken around the world. This has deepened and clarified early ideas, consolidated the role of adaptive comfort and informed this book, not least through the Windsor Conferences which they initiated and which continue until today.

In this book we describe the adaptive approach to thermal comfort. We present some of the latest work in the field and suggest the ways in which it is being developed into a model. Such a model can be used to set dynamic, interactive standards for thermal comfort which will help overcome the problems inherited from the past. The approach we present is by no means the last word on the subject. Indeed for any small group of researchers to claim hegemony in such an undertaking would be foolish as well as arrogant. We try to present our own understanding of the processes of adaptive comfort and their implications for the design of better buildings – and we look forward to any discussions this may provoke!

The book is divided into two sections. The first introduces the ideas behind the study of thermal comfort and in particular the adaptive approach. It tries to show the ways in which the adaptive model would lead to improvements in comfort in buildings at the same time as reducing the energy they use. The second section outlines practical ways in which to go about thermal comfort work in the field.

This volume provides a scientific methodology for field studies to parallel the methodology that underpins laboratory-based models. The method is based on work in the field among populations familiar with the conditions being measured, and must be widely applied if the resulting adaptive model is to become the standard model. This can only be achieved through the worldwide effort, started by Michael Humphreys and Fergus Nicol in the 1970s and continued around the world by Andres Auliciems, Richard de Dear, Gail Brager, Nick Baker, Nigel Oseland and a growing band of researchers, in an attempt to map comfort conditions in every major climatic zone. No single team is able to undertake such a task without the active help of numerous local groups of researchers in countries across the whole world.

One purpose of this book is to lay down the theoretical and practical bases for the design and implementation of such field surveys. We do not seek to 'tell anyone what to do', but more to lay down a structured foundation of knowledge from which the low carbon, comfortable buildings of the future can be built. We hope that local researchers will use the contents of these volumes to help them in the conduct and analysis of their own field surveys, and that they will in their turn make their own findings available to others interested in the development of a better understanding of thermal comfort in buildings.

Finally we should say something about the style of this book. Remembering that our readers will be from a variety of backgrounds and academic disciplines – engineering, architecture and the sciences – we have tried to make the book accessible to all by avoiding unnecessary use of technical vocabulary and jargon. Sometimes technical vocabulary is essential, and where this is so we have explained the terms we use. The combination of simplicity and precision is difficult to achieve, and no doubt we have sometimes strayed to one side or the other. The formal impersonal style so prevalent in the sciences obscures the human element in social and scientific research. We have preferred to address you, our readers, more directly, because adaptive thermal comfort is about people, and field research is undertaken by people, with people and for people like us.

# References

Auliciems, A. (1990) Air conditioning in Australia III: Thermobile controls, *Architectural Science Review* 33(2), 43–48.

Auliciems, A. and de Dear, R.J. (1986) Air-conditioning in Australia part I. Human thermal factors, *Architectural Science Review* 29, 67–75.

Langdon, F.J., Humphreys, M.A. and Nicol, J.F. (eds) (1973) *Thermal Comfort and Moderate Heat Stress*. London: HMSO.

Mendell, M.J. and Mirer, A.G. (2009) Indoor thermal factors and symptoms in office workers: Findings from the US EPA BASE study, *Indoor Air* 19, 291–302, Blackwell Publications.

Nicol, J.F. and Humphreys, M.A. (1973) Thermal comfort as part of a self-regulating system. *Building Research and Practice* (J. CIB) 6(3), 191–197.

# ACKNOWLEDGEMENTS

This is the first book in our trilogy on adaptive thermal comfort. The lead author on Volume 1 is Fergus Nicol, and this book covers the fundamental theory underpinning the subject and describes the related methods, calculations, field survey strategies and analysis methods for researchers and building designers. Volume 2, led by Michael Humphreys, covers the advanced theory of adaptive comfort and Volume 3, led by Sue Roaf, deals with how to design low-carbon buildings using adaptive thermal comfort strategies.

The first version of Volume 1 (called *Thermal Comfort: A Handbook for Field Surveys Toward an Adaptive Model*) was written by Fergus Nicol and came out as a temporary publication at the instigation of Mike Thompson at the University of East London in 1993. It was specifically designed to encourage the promotion of field surveys in thermal comfort teaching and research, and to use the adaptive approach to analyse them. Marialena Nikolopoulou sponsored a lightly edited and updated second draft, which was published in electronic format by the University of Bath in 2008 as *A Handbook of Adaptive Thermal Comfort: Towards a Dynamic Model*. Both versions have been frequently cited but the subject has moved on rapidly, so this volume is not only a comprehensive revision of the first two versions but also includes new chapters and thinking on the issues and methods involved.

Many colleagues at Oxford Brookes, East London, London Metropolitan and Heriot-Watt Universities have given particular cooperation, support and encouragement to the authors, including Iftikhar Raja, Hom Rijal, Kate Dapré (formerly McCartney), Mary Hancock, Maita Kessler, Jo Saady, Ollie Sykes, Jane Matthews, Mike Wilson, John Solomon, Phil Wilkins, Axel Jacobs, Luisa Brotas, Janet Rudge, Laia Cunill and Jo Dubiel.

We are particularly indebted to Richard de Dear, Gail Brager, Nick Baker, Ken Parsons, Bjarne Olesen and Atze Boerstra for insights and collaborations. Colleagues with whom we have had the benefit of working on international projects include Arif Allaudin and Gul Najam Jamy in Pakistan; Gary Clark, Paul Langford, Mick Hutchins, Paul Tuohy and Aizaz Samuel in the UK; Jan Brissman and John Stoops (now in the US) in Sweden; Mat Santamouris, Aris Tsangrassoulis, Costas Pavlou, Katcrina Sfakianaki from Greece; Eduardo Maldonado, Jose Luis Alexandre, Alexandre Friere from Portugal; Pierre Michel, Gerard Guarracino, Francis Allard and Cristian Gauius from France; Lorenzo Pagliano and Paulo

Zangeri from Italy; Wilfried Pohl from Austria; and Alison Kwok in the US; and Chiheb Bouden and Nadia Ghrab-Marcos in Tunisia.

Among the wider fraternity of thermal comfort research we owe a debt of gratitude to Margaret Gidman, Adrian Pitts, Steve Sharples, Bill Bordass, George Havenith, Gary Raw and Nigel Oseland in the UK; Arvind Krishan and Ashok Lall in India; Tri Harso Karyono in Indonesia; Martin Evans in Argentina; Roberto Lamberts in Brazil; Shahin Heidari and Omid Saberi in Iran; Terry Williamson, Bruce Forewood and Steve Szokolay in Australia; Ed Arens, Hui Zhang, Alison Kwok and Gregor Henze in the US; Jens Pfafferott, Doreen Kalz, Gerd Jendritsky and Dusan Fiala in Germany; Stanley Kurvers in Holland; and Yufeng Zhang, Baizhan Li, Edward Ng and Jiang Yi in China.

One of the joys of research work in thermal comfort is the international nature of the effort. Strong interest and support have been forthcoming from all corners of the globe with particular contributions from Japan, Brazil, Australia, the Far East, the US and throughout Europe. Most recently there has been an upsurge of interest from China and an emerging group in Africa. Our thanks to all the colleagues from around the world who have regularly joined us at the Windsor Conferences to share ideas, discuss ways forward and to work together to refine and advance thermal comfort research. They have helped us to keep us abreast of the evolving comfort research needs, trends and priorities in a rapidly changing world.

Our students at Oxford Brookes, East London, London Metropolitan and Heriot-Watt Universities are always a source of motivation and inspiration.

Finally we extend our thanks to Gill Nicol and Mary Humphreys for reading and improving various drafts and to Rex and Jane Galbraith for valuable statistical advice. Nicki Dennis, of Earthscan and Routledge had the courage to invite us to produce three books on this rather specialist subject, a decision for which we remain very grateful.

Permission to reproduce extracts from BS EN ISO 7726, 7730, 8996 and 9920 is granted by BSI. British standards can be obtained in PDF or hard copy formats from the BSI online shop: www.bsigroup.com/shop or by contacting customer services for hard copies only: tel +44 (0)20 8996 9001, email cservices@bsigroup.com.

Needless to say, responsibility for errors or omissions remains our own.

Fergus Nicol, Michael Humphreys and Susan Roaf

The UK Collaborative Group on Thermal Comfort, which backed the first edition of this book, has since developed into a wide-ranging but informal group of thermal comfort researchers from all disciplines. The Network for Comfort and Energy Use in Buildings (NCEUB) holds seminars and meetings at which members bring each other up to date on their latest research and exchange ideas and experiences. NCEUB is also responsible for the prestigious international Windsor Conferences held at the Cumberland Lodge conference centre in Windsor Great Park. Almost all of the papers given at the Windsor Conference in the past ten years are available on our website as well as presentations, meeting reports and the work of our members. This is a free resource for our members and for the public. If you want to get involved visit our website – http://nceub.org.uk.

# ABBREVIATIONS

| | |
|---|---|
| ANSI | American National Standards Institution |
| ASHRAE | American Society of Heating, Refrigerating and Air Conditioning Engineers |
| BRE | Building Research Establishment |
| CEN | Comité Européen de Normalisation |
| CIBSE | Chartered Institution of Building Services Engineers |
| EPBD | Energy Performance of Buildings Directive |
| HSE | Health and Safety Executive |
| HVAC | Heating, Ventilating and Air Conditioning |
| ISO | International Standards Organisation |
| LED | light emitting diode |
| MRT | mean radiant temperature |
| NCEUB | Network for Comfort and Energy Use in Buildings |
| PLEA | Passive and Low Energy Architecture |
| PMV | predicted mean vote |
| POE | post-occupancy evaluation |
| PPD | predicted percentage dissatisfied |
| RH | relative humidity |
| SET | standard effective temperature |
| TES | thermal energy storage |

# PART I

# Principles
Building an adaptive model

# 1

# THERMAL COMFORT

## Why it is important

Most people have a daily thermal routine. Within limits we expect to know how warm the bedroom will be when we awake, the kitchen when we have breakfast. We also know what to expect on the bus (or bicycle) journey to work and at the office when we arrive. These 'expected' environments can vary from time to time and from season to season, but on the whole we know what thermal conditions to expect over a day or month and we will generally have strategies for dealing with them – or rushing through those parts that are not acceptable. So why is it important to know which temperatures are comfortable? What is the purpose of the science of thermal comfort?

## 1.1 User satisfaction

### 1.1.1 Comfort

In his survey of user satisfaction in buildings with passive solar features, Griffiths (1990) found that having the 'right temperature' was one of the things people considered most important about a building. This is a result that has been borne out by many other surveys over the years. Griffiths also found that 'air freshness' was an important requirement mentioned by the respondents to his survey. The subjective freshness of the air was found by Croome and Gan (1994) among others to be closely related to the temperature of the air. In other words, two important features in the user satisfaction with a building are closely related to temperature.

At the same time, dissatisfaction with the thermal environment is widespread, even in buildings with sophisticated controls. Complaints of overheating in winter and coldness in air-conditioned buildings in summertime are commonplace. A survey in an air-conditioned building carried out by students of Oxford Brookes University with the Building Research Establishment (BRE) found almost 30 per cent of occupants found the building too hot in winter.

### 1.1.2 Health

We are mammals and the human body must be kept at a constant internal temperature known as our 'core temperature'. Much of the time this is done by physiological processes, by diverting the blood towards or away from the periphery of the body to control the heat loss from the skin, by producing sweat to cool the skin by evaporation, or by shivering when we are cold. In this context thermal discomfort is essentially a warning that the environment might present a danger to health. If we feel too cold we will take actions to make ourselves warm and comfortable again, or if too warm, to cool ourselves before the core temperature changes too much.

The threat of climate change and the increasing instances of hotter than normal conditions reinforces the pressing need (Nakicenovic and Swart, 2010; Figure 1.1 in Plates) to improve the buildings we occupy to protect ourselves from future climate conditions:

> The spike … is the 2003 heatwave that killed about 35,000 people in France, Italy and Spain. The continuation of the heating trend under mid-range climate change scenarios would make the heatwave – which was about a one-in-a-100-year event at the time it occurred and a one-in-250-year event before humans started fiddling with climate – into a one-in-two-year event by 2050. In 2070, those deadly conditions of 2003 will be considered an unusually cool summer.
>
> (Holdren, 2008)

Cold winters continue to result in excess winter deaths not least because an increasing number of people are falling into fuel poverty as the cost of gas and electricity soar and the global recession takes its toll. A person or family is said to be in fuel poverty in the UK if more than 10 per cent of their disposable income is required to ensure a comfortable temperature in their home. The number of people in fuel poverty is thus a function of the cost of fuel as well as the incidence of poverty in the population. Rising fuel poverty is characterised by the rise in deaths and heat- and cold-related illnesses become more common causing a rise in morbidity (illnesses) and mortality (deaths) during extreme weather events (Roaf et al., 2009). Even in many developed nations fuel poverty levels are very high. By the summer of 2011 around 35 per cent of all homes in Scotland were deemed to be in fuel poverty.

### 1.1.3 Delight

In her book *Thermal Delight in Architecture*, Lisa Heschong (1979) suggests that we can and should look for more than simply comfort in buildings, and consider that our thermal sense is capable of giving not just 'satisfaction' but in certain circumstances actually producing delight. This is especially true when we experience a variable environment. Following the work of Chatonnet and Cabanac (1965) and Cabanac (1992) on the ways in which our thermal sensations relate to comfort, Richard de Dear (2011) relates the delight we feel in the thermal environment in those aspects that tend to return the thermal state of the body to equilibrium. So the movement of the air around us, which tends to cool the body, is unpleasant when we are already feeling cold but can be delightful if we are warm.

## 1.2 Energy consumption

The indoor temperature that is set for a building in the heating or cooling season is key to the energy used in the building. The heat loss from, or gain in, the building depends on the indoor–outdoor temperature difference. In the UK roughly 10 per cent of the heating energy used in winter is typically saved by a reduction of 1K in the indoor temperature.[1] This saving includes contributions from the reduced indoor–outdoor temperature difference and a decrease in the length of the heating season. A lower indoor temperature means that the heating can be turned on later in the autumn, and can be turned off earlier in the spring. Because of a reduction in maximum heat load the buildings will also need a less powerful heating system.

Similar considerations apply to air-conditioned buildings: a decrease in the outdoor–indoor temperature difference will decrease the 'cooling load' – the energy needed to cool the building to a comfortable temperature. There are two factors that make the reduction of cooling loads for air conditioning an area of critical concern. First, air conditioning uses electrical power, which is highly inefficient in its generation if fossil fuels are used. It consequently wastes large amounts of energy to cool a building. Haves *et al.* (1992) have suggested that the energy used and wasted in the air conditioning of buildings is a significant driver for global warming and creates a positive feedback loop because higher temperatures result in more air conditioning use.

Second, the wide-scale use of air conditioning in countries such as the USA has led to 'energy hog' buildings and has discouraged buildings that use 'passive' – non mechanical – cooling strategies (Cooper, 1998; Ackerman, 2002). The need for air conditioning can often be significantly reduced or removed altogether by improving the thermal performance of the building through strategies such as reduced glazed areas, more shading, increasing levels of thermal mass, and naturally ventilating the building for as much of the year as possible.

There is no question that in some parts of the world buildings require heating, and in others parts they need cooling to remain habitable (Figure 1.2). Our challenge is to minimise the period of the year over which these systems need to be used. The two key elements in the solution to this challenge are to design better buildings and to use an adaptive approach to achieving comfort in them.

## 1.3 Standards, guidelines and legislation for indoor temperature

Both the comfort of the occupants of a building and the energy it consumes are closely linked to indoor temperature. The indoor temperature is clearly a prime concern for the owners of buildings and the people who live and work in them. The method by which we decide what temperature to aim for in a building therefore has far-reaching consequences.

One way in which we can decide what temperature to provide in a building is by reference to international temperature standards. Most temperature standards suggest a method

---

1  A simple method used to assess heating requirements for buildings is to find, from weather data, the number of 'heating degree-days' in the heating season. This is found by multiplying the days during which there is an indoor–outdoor temperature drop by the number of degrees temperature drop for each day. In the UK, for instance, the number of design degree-days for an assumed indoor temperature of 18.0°C is about 2000, for 15.5°C it is about 1500. These figures indicate a drop of 25 per cent in the heating energy used for a drop of 2.5K in the internal temperature – or a saving of about 10 per cent for every degree.

for deciding an optimal temperature for a given space or building depending on its intended use, especially if the temperature is mechanically regulated by the heating or cooling system. ISO 7730 (2005) is an example of such a standard. The temperatures recommended result from calculations based on the activity of the occupants of a building and their clothing (more will be said about this approach in Chapter 4).

Guidelines for deciding on the best indoor temperature for different types of buildings or for different spaces within a building are available from such sources as CIBSE Guide A (2006) or the ASHRAE Guide and Databook (2009). It is to these sources that many heating and ventilating engineers will turn when deciding what temperatures to use in their calculations. These guidelines are based on professional experience and informed by comfort standards but reflect the assumption that for a given activity there is a given 'best temperature' and that this is correct for all circumstances. We know of no similar source for architects and this may be a reflection of the way they have handed responsibility for the indoor environment to engineers, to the detriment of the architectural profession and possibly of the buildings they design.

In most countries there is also legislation that sets limits on temperature, with the aim of protecting health and well-being, such as the UK Offices, Shops and Railway Premises Act (1963) which sets a minimum temperature for working environments.

## 1.4 Adaptation

Experience tells us that an indoor temperature that varies with the weather is possible and often desirable in any building. Temperatures that are 'right' in summer when wearing a T-shirt and shorts could be oppressively hot in winter clothing. We expect different thermal experiences in summer and winter, and we modify our behaviour accordingly. The relationship between indoor comfort and outdoor temperatures in different climates can be used to suggest appropriate features for a low-energy design. This can be done by using the Nicol graph (Roaf et al., 2012) which relates preferred indoor comfort temperatures of adapted populations to outdoor temperatures at different times of year for a given climate (see Figure 6.4).

Recently international standards (e.g. ASHRAE 55-2004) and the European Standard EN15251 (CEN, 2007) have recognised the possibility that the comfort temperature can vary with changing outdoor conditions. These variable or adaptive standards are taken to apply in buildings that are naturally ventilated. A similar relationship could be assumed for heated or cooled (mechanically conditioned) buildings and certainly no validated reasons have been suggested why this should not be so. A variable standard requires comfort temperatures to change with the climate surrounding the building. It reduces the average indoor–outdoor temperature difference, and consequently reduces energy requirements considerably as against a single-temperature standard. Some authors suggest this could be by as much as 50 per cent of the total heating and cooling requirements for a building (Auliciems, 1990) but these estimates may not take account of the importance of fan power in the cooling loads.

The Heating, Ventilating and Air Conditioning (HVAC) industry can control the indoor environment mechanically to supply a single constant temperature all year round. However, the need to reduce energy use in buildings suggests that where possible buildings should be designed to deliver comfortable indoor conditions without the use of energy.

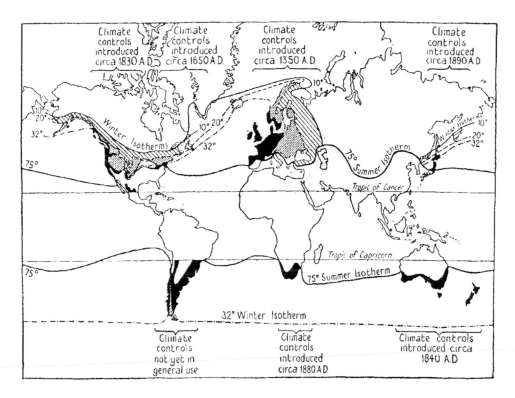

**FIGURE 1.2** A map of the world in 1939 by S.F. Markham showing those parts of the world in black that are most suitable for habitation by Homo Sapiens because they are the areas that require neither excessive heating nor cooling of buildings. Markham was clear that the miracle of air conditioning would open large parts of the world to exploitation by European white men who function best in cooler temperatures.

Because the current common standards relate to buildings with mechanical conditioning they seem to expect that all buildings will provide uniform indoor environments. Traditional buildings provide a variety of thermal environments between rooms and within a single room. The bedroom in a traditional English house is cooler than the living room in winter, but the bed is warm and snug. A coal-burning fire gives a focus of radiant heat that provides a variety of environments depending on how close you are to it and so on. In the Middle East, large traditional houses owned by the wealthy have many rooms each with its own particular thermal characteristics, and the inhabitants could choose the most appropriate climate in the complex of spaces at different times of day and year.

The relationship between people and environment is complex and active, bringing in time, climate, building form, social conditioning, economic and other factors as well as the immediate physical environment. This complexity implies that the indoor environment needs to take this variability into account. This means an indoor environment that changes with the season and the climate, that allows buildings to change, suggests how quickly they should do so and reflects the willingness of occupants to vary their environment by returning some measure of control to them.

In this book we aim to describe the system within which human thermal requirements operate. An understanding of this system can be used to explain and quantify the interactions between man and environment, and in turn build an improved specification of comfortable temperatures in buildings.

The adaptive approach to comfort here is based on the Adaptive Principle:

*If a change occurs such as to produce discomfort, people react in ways which tend to restore their comfort.*

The search for thermal comfort arises from the need of our bodies to maintain a stable core body temperature. The move towards comfort is part of a dynamic process exhibited partly through the interaction between people and buildings that is in turn dependent on economic and social conditions as well as the thermal environment. Our ability to be comfortable is affected by larger factors such as our exposure to uncomfortable conditions and climates, our personal profile such as health, wealth, age and the way we manage our lives. A key factor is also the access we have to the means and opportunities to control our environment and the constraints imposed on our ability to use them, be it in the form of strict clothing and management rules at work or a complete disconnect between building user and the building management systems in an open-plan, fixed-window working environment.

The complex interactions involved in the larger comfort system cannot be fully explored in the laboratory. That is why the adaptive approach is built upon the results of surveys in the field, with people making thermal choices in the context of their own comfort systems. This book aims to set the ideas behind adaptive comfort in context, to describe those systems and to suggest the ways in which thermal adaptation can be explored and comfort achieved in our complex world.

## 1.5 Comfort outdoors and in intermediate spaces

There is increasing interest in providing comfort not just indoors but also in intermediate and outdoor spaces. Here again an adaptive approach is appropriate and much can be gained from field studies. Adaptive opportunities that apply indoors, such as the opportunity to use windows and heaters, are not applicable in an outdoor context, but providing shade from the sun or a shield from the wind will amount to providing adaptive opportunities. Of course in outdoor conditions clothing choice will provide a major opportunity for comfort (Humphreys, 1979), though field surveys have found that complete adaptation to the weather is not generally achieved, so that people outdoors are generally cold in cold weather and warm in hot weather (Nicol *et al.* 2006). This incomplete adaptation may be because in general people are only out of doors for relatively short periods of time unless they have an outdoor job.

Intermediate spaces such as entrance hallways or canopies are also important in allowing people who have been outdoors to become accustomed to indoor conditions and can be used to reduce energy use and improve people's comfort (Bin Saleh and Pitts, 2006).

# References

Ackerman, M. (2002) *Cool Comfort: America's romance with air-conditioning*. Washington, DC: Smithsonian Institution Press.

ASHRAE (2004) *ANSI/ASHRAE Standard 55-2004: Thermal environmental conditions for human occupancy*. Atlanta, Georgia: American Society of Heating, Refrigerating and Air Conditioning Engineers.

ASHRAE (2009) *ASHRAE handbook of fundamentals*. Atlanta, Georgia: American Society of Heating, Refrigerating and Air Conditioning Engineers.

Auliciems, A. (1990) Air conditioning in Australia III: Thermobile controls, *Architectural Science Review* 33(2), 43–48.

Bin Saleh, J. and Pitts, A. (2006) Potential for energy saving in transition spaces, *Proceedings of the Fourth Windsor Conference: Comfort and Energy Use in Buildings: Getting it right, 27–30 April 2006*. London: Network for Comfort and Energy Use in Buildings (NCEUB). Available at http://nceub.org.uk.

Cabanac, M. (1992) What is sensation? In R. Wong (ed.) *Biological Perspectives on Motivated Activities*. Northwood, NJ: Ablex.

CEN (2007) Standard EN15251. *Indoor Environmental Parameters for Design and Assessment of Energy Performance of Buildings: Addressing indoor air quality, thermal environment, lighting and acoustics*. Brussels: Comité Européen de Normalisation.

Chatonnet, J. and Cabanac, M. (1965) The perception of thermal comfort. *International Journal of Biometeorology* 9(2), 183–193.

CIBSE (2006) Chapter 1, Environmental criteria for design, in *Environmental Design: CIBSE Guide A*. London: Chartered Institution of Building Services Engineers.

Cooper, G. (1998) *Air-conditioning America: Engineers and the controlled environment, 1900–1960*. Johns Hopkins Studies in the History of Technology. Baltimore, MD: John Hopkins University Press.

Croome, D.J. and Gan, G. (1994) Thermal comfort and air quality in naturally ventilated offices: Case study, *Building Services Research and Technology* 15(3), 27–28.

de Dear, R.J. (2011) Revisiting an old hypothesis of human thermal perception: Alliesthesia, *Building Research and Information* 39(2), 108–117.

Griffiths, I. (1990) Thermal comfort studies in buildings with passive solar features: Field studies: *Report to the Commission of the European Community, ENS35 090 UK*.

Haves, P., Kenny, G. and Roaf, S. (1992) *The impact of climate change on buildings*, Conference of Proc Passive and Low Energy Architecture, Auckland.

Heschong, L. (1979) *Thermal Delight in Architecture*. Cambridge, MA: MIT Press.

Holdren, J.P. (2008) *Meeting the Climate-Change Challenge*. Eighth Annual John H. Chafee Memorial Lecture on Science and the Environment, National Council for Science and the Environment, Washington. Available at www.NCSEonline.org.

Humphreys, M.A. (1979) The influence of season and ambient temperature on human clothing behaviour. In P.O. Fanger and O. Valbjorn (eds) *Indoor Climate*. Copenhagen: Danish Building Research.

ISO TT30 (2005) *International Standard 7730 Moderate thermal environments – determination of the PMV and PPD indices and specification of the conditions for thermal comfort*. Geneva: International Standards Organisation.

Nakicenovic, N. and Swart, R. (2010) *Special Report on Emission Scenarios*. Cambridge: Cambridge University Press.

Nicol, F., Wilson, E., Ueberjahn-Tritta, A., Nanayakkara, L. and Kessler, M. (2006) Comfort in outdoor spaces in Manchester and Lewes, UK, *Proceedings of the Fourth Windsor Conference: Comfort and Energy Use in Buildings: Getting it right, 27–30 April 2006*. London: Network for Comfort and Energy Use in Buildings. Available at http://nceub.org.uk.

*Offices Shops and Railway Premises Act* 1963. London: HMSO.

Roaf, S., Crichton, D. and Nicol, F. (2009) *Adapting Buildings and Cities for Climate Change*. Oxford: Architectural Press (see chapter 10 on Health and Climate with contributions by Sari Kovats and Janet Rudge).

Roaf, S., Fuentes, M. and Thomas, S. (2012) *Ecohouse: A Design Guide* (4th edn). Oxford: Architectural Press.

# 2

# THERMAL COMFORT

## The underlying processes

Our thermal comfort is linked to the need to maintain an almost constant internal temperature, irrespective of the amount of heat we produce within our bodies or what environment we are in. This stable core body temperature of around 37°C is essential for our health and well-being. Our thermal interaction with the environment is directed towards maintaining this stability in a process called 'thermoregulation'. The opportunities available to us for such interactions are numerous and complex and are the subject of a great deal of research. Thermal physiologists study how we produce and use heat. Psychologists interpret our conscious feelings about the environment. The way in which heat is transferred from our body to the environment is studied by physicists. Sociologists analyse the way we react to the environment. Finally it is the role of the architect or building engineer to design buildings that best meet our thermal needs. Those who study thermal comfort have to deal with all these factors and disciplines.

## 2.1 Physiology

Most of the food we eat is converted into heat that the body produces all the time. The more active we are the more heat we produce. Muscular activity is most associated with heat production, but all bodily functions, such as digesting food or even thinking, produce some heat. The amount of heat produced is usually expressed in watts (W) per square metre (m²) of body surface area. A reclining person may produce 60 W/m² while a sprinter may be generating over 400 W/m² of body area. The skin surface area of an average adult is about 1.7m². Heat is transported around the body in the blood, and is then transferred by conduction through the body tissues from the warm interior to the cooler periphery of the skin and on into the environment, resulting in heat loss from the body. If the external air temperature is above the mean skin temperature then the process is reversed, causing heat to flow from the air into the body. The heat produced in the body by the metabolic processes must, over a period, equal the heat it loses to the environment.

The temperature of the internal organs (the 'deep-body' or 'core' temperature) must be maintained within close limits for the proper functioning of the organs of the body, and particularly the brain. The seat of the controlling mechanism is within the brain. If our brain temperature goes outside the limits, the body will react physiologically to restore its proper temperature.

If the body temperature drops, vasoconstriction is initiated. Blood circulation to the peripheral parts of the body (most noticeably the hands and feet) is reduced through the contraction of the surface blood vessels. The consequent reduction in the supply of metabolic heat to the skin causes its temperature to drop. This reduces the overall rate of heat loss from the skin to the surroundings, so conserving valuable heat internally to help the body maintain its core temperature. Any further drop in core temperature leads first to an increase in the tension in the muscles and then to shivering; both of these processes increase metabolic heat production.

If the core body temperature rises the first line of defence is vasodilation: the blood vessels close to the skin expand to increase blood supply to the periphery, flushing the skin with blood. This increases the skin temperature and consequently the rate of heat loss to the surroundings. Further increases in core temperature give rise to sweating. This causes heat loss when sweat is evaporated from the surface, absorbing heat from the adjacent skin and air.

Much of the work done by thermal physiologists has been concerned with the limits of human endurance in extreme conditions of heat and cold stress as may be experienced by mountaineers, steel factory workers, coal miners and soldiers on active duty in extreme climates. However, in ordinary buildings we are typically dealing with no more than moderate heat and cold stress, although in recent years the numbers of people actually dying of heat and cold stress in ordinary buildings have risen as more extreme weather events become more common in the changing climate.

## 2.2 Psychophysics

The unconscious thermoregulatory actions controlled from the brain are augmented by our ability to sense our thermal environment, particularly through the skin. Our thermal sense gives us information about skin temperature, providing warning of conditions that might pose a danger to the internal core temperature. The way in which the skin sensors act is time-dependent, so that a sudden change in skin temperature gives an immediate steep rise in sensation which then dies away (Cabanac, 1992). A slower rise in skin temperature gives a more gradual increase in thermal sensation and may mean that people underestimate the change.

Our impression of the warmth or coldness of our environment arises in part from these skin sensors. The brain coordinates the response to the thermal landscape of our whole body and is centred on the need to maintain the core body temperature. So the overall sensation we feel will be pleasing or displeasing depending on whether the overall effect of our local conditions results in us moving towards, or away from, the restoration of deep body thermal equilibrium. Thus a cold sensation will be pleasing when the body is overheated but unpleasant if the core is already cold. Similarly, a hot sensation is pleasant if we are cold, but oppressive if we are already hot. At the same time, the temperature of the skin is by no means uniform. As well as general variations caused by changes in metabolic rate, there are wide variations in the surface temperatures of different parts of the body caused by how close the blood supply is to the skin, how good the local circulation system is and how much body fat there is. Fingers and toes tend to be the coldest parts of the body surface and armpits among the warmest. As might be expected, clothes have a marked effect on the level and distribution of skin temperature. So sensation in any particular part of the skin will depend on many different variables including time, clothing and the temperature of the surroundings.

Researchers have identified many factors that contribute to people experiencing a particular sensation. The recent thermal history of a person is key. Have they just finished playing tennis? Their age, sex, culture, personality type and expectation of what they are about to do, may all influence the way they feel. In the longer term, people's familiarity with the conditions they encounter is important. Hot dry climates require that people drink a lot of water to maintain thermal balance. People who are from a different climate or are subject to a sudden change in weather may not recognise this need and run the danger of heat stroke.

Psychophysics is the study of the relation between our sensations, the stimuli we receive from the physical world and the ways in which our brain interprets them. The science of psychophysics has been concerned with investigating the relationship between sensations such as apparent brightness and stimulus of the light level, or the sensation of loudness and the stimulus of sound level. For a good introduction to psychophysics applied to the built environment see Hopkinson (1963). The relationship between overall thermal sensation and the characteristics of the environment is hard to pin down. Psychophysical experiments have tended to concentrate on the more defined relationships such as that of the perception of the warmth and cold of surfaces when they are touched. Nevertheless, the thinking behind many studies of thermal comfort has been that of psychophysics: relating the overall thermal sensation to the stimuli of the thermal environment (see Chapter 3).

Psychophysicists have been able to demonstrate that we can assign a number or description to our sensations. In our response to the thermal environment we act like a 'sensation-thermometer'. Many researchers have tried to construct indices that will mimic this human response to the environment (Auliciems and Szokolay, 2007) and this is a subject we return to in Chapter 4. The number of distinct sensations we can reliably distinguish is limited, and the normal scale of thermal comfort, having seven or so descriptive points (Table 2.1), recognises this (Miller, 1956).

The ability to say how we feel does not imply a one-to-one relationship between comfort and the physical conditions causing it. Just as we have shown that the pleasure given by a thermal stimulus depends on the physiological conditions against which the stimulus is received, so uncertainty about the thermal sensation is caused not just by the physiological but also by the social and cultural context. People can be comfortable in a range of circumstances. So we can't say that a particular set of conditions will give rise to such-and-such a sensation, only that there is a certain probability that it will. A person brought up in a Bedouin tent in Iraq will obviously experience a temperature very differently from a person brought up in a Sani tent in the Russian Arctic. But a person who lives and works in an air-conditioned building in Maceio city in north-eastern Brazil also experiences temperatures differently from his neighbour who lives and works in naturally ventilated buildings (Cândido et al., 2010).

## 2.3 Physics

To the physicist, the human being is a heated body with variable surface characteristics, losing most of its heat to the environment in three ways – convection, radiation and evaporation. In certain circumstances some heat is also lost from the soles of our shoes or clothing surfaces by conduction to cold surfaces they are in contact with, such as a floor we stand on or a stone bench we sit on, but this is not often an important factor in our day-to-day experience. The amount of metabolic heat produced by the body when lightly active is quite small in relation to the mass of our bodies, and the difference between the heat input from metabolism and the heat loss to the environment is usually even smaller, so changes in our thermal state can take

**TABLE 2.1** Two seven-point scales commonly used in thermal comfort work

| ASHRAE* scale | | Bedford scale | |
|---|---|---|---|
| Descriptor | Number | Descriptor | Number |
| Hot | + 3 | Much too warm | 7 |
| Warm | + 2 | Too warm | 6 |
| Slightly warm | + 1 | Comfortably warm | 5 |
| Neutral | 0 | Comfortable neither warm nor cool | 4 |
| Slightly cool | − 1 | Comfortably cool | 3 |
| Cool | − 2 | Too cool | 2 |
| Cold | − 3 | Much too cool | 1 |

*American Society of Heating, Refrigerating and Air Conditioning Engineers

time to happen. A light person will respond more quickly than a heavy person, which may explain why a thin person feels the cold more quickly than someone who is heavily built.

Some of the energy produced by the body is used to do work. A person sawing wood will expend extra energy while sawing. Some of this energy will end up as heat produced by the cutting process itself, but this is external to the body so has to be deducted from metabolic heat when considering its energy balance.

Over any substantial period the heat produced within the body must balance the heat lost from the body. This is done through the three physical processes: convection (heat lost through the body heating the air around us); radiation (heat radiated to surrounding surfaces); and evaporation (heat lost by evaporating sweat and other moisture). Some of the heat loss by convection and evaporation is through respiration and this is generally considered separately. The mathematical expression of this and other processes are given in Section 2.5.

### 2.3.1 Radiation

Any body emits radiant heat from its surface at a rate that is proportional to the fourth power of its absolute temperature. The absolute temperature is equal to the Celsius temperature plus 273, relating it to 'absolute zero' at which no radiation is emitted. At the same time surrounding surfaces are radiating to the body in a similar way. So a balance is set up whereby the body will lose heat if the surroundings are colder and gain heat if they are hotter. If the surroundings were all at one temperature the situation would be relatively simple. But often they are not; in a real situation there can be cold windows or hot ceilings, hot radiators and even radiation from the sun.

The radiation from heated surfaces in a room is generally long-wave or infra-red radiation and is invisible (Figure 2.1 in Plates). From very hot surfaces some of the radiation becomes visible and is referred to as short-wave or visible radiation (Figure 2.2). Visible radiation also contains heat. About half of the radiant energy from the sun is emitted in the visible range. From an incandescent light bulb it is much less. This means that solar radiation is relatively efficient as a light source. More efficient light sources such as fluorescents or light emitting diodes (LEDs) give out less heat but the light energy will still cause some radiant heating. This will be part of the radiant heat 'landscape' of a room.

Quantifying the overall effect of the full radiant landscape of a room would involve the integration of the effect of every part of the surrounding surfaces and other radiant sources with every part of the body. The usual simplification of this complex matter is to evaluate the 'mean radiant temperature' of a space. This may be defined as the temperature of a small sphere centred at the point in question which would exchange no net radiation with the surroundings because the incoming and outgoing radiation were equal. The radiation actually exchanged with the surroundings will then be proportional to the difference between the fourth power of this imaginary temperature and that of the actual temperature of the sphere. The contribution of any particular surface to the mean radiant temperature at a point depends on its temperature, its surface characteristics and the solid angle it subtends at the point in question. The mean radiant temperature might seem an artificial simplification, because people are not 'small spheres', but the notion of the mean radiant temperature is very useful.

The mean radiant temperature can vary from point to point in a room. In many instances a rough estimate of radiant temperature such as the average temperature of the room surfaces is all that can be achieved. Remember that we are talking here of the surfaces that can be 'seen' from the position of the person and in a furnished room these may be overwhelmingly the surfaces of the furnishing rather than the wall, ceiling or floor surfaces of the room. Lightweight furnishing surfaces are often close to air temperature and consequently the mean radiant temperature in a room is typically close to air temperature. The effect of radiation heat loss to a large cold surface such as a window does still need to be taken into account especially for the people close to it. People notice the resulting cooling and often interpret it wrongly as a 'draught'.

Because there is only a relatively small temperature range in human environments, a simple approximation to the fourth power law for radiant heat exchange can be used. The approximation takes the radiant heat loss to be proportional to the temperature difference between the surface of the clothed body and the mean radiant temperature.

### 2.3.2 Convection

Our body is surrounded by air with which it exchanges heat, largely at the surface of the skin (or clothes) and the surfaces in the lungs. If the temperature of the air is lower than that of the skin or the surface of the clothes there will be a heat loss from the body by convection as heat is carried away by the air. In cool, still conditions air movement around the body will be caused by the air, heated by the body, rising up over it forming a 'plume' above the head and dispersing. This process is known as 'natural convection' (Figure 2.3 in Plates).

Any additional air movement around the body will add to the cooling by helping to strip away the warmed air close to the skin at a greater rate. The air movement is relative to the body surface, so activities such as walking will add to the air movement.

Convective heat loss will occur whether or not the body is clothed. When the body is clothed, the cooling effect of convection occurs mostly at the outer surface of the clothing. The overall effect of convection is to lower the temperature of the clothing surface, indirectly

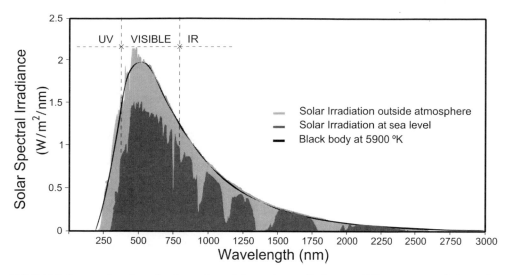

**FIGURE 2.2** Image showing the intensity of incoming radiation at different parts of the solar spectrum

Source: Mick Hutchins.

cooling the skin. Air movement can reduce the effectiveness of clothing as an insulator, as will the wearing of loose clothing that swirls air over the skin as it moves.

Where the air in contact with the body is hotter than the skin (usually around 32°C to 35°C), more exposure to moving air will act in the opposite direction, heating the body rather than cooling it. At these higher temperatures, above 35°C, evaporative cooling becomes the key thermoregulatory strategy for cooling.

In conclusion, the cooling (or heating) effect of the air depends on the difference between air temperature and skin temperature (or clothing surface temperature) and on the speed of the air movement. The cooling effect of air movement has been found to be roughly proportional to the square root of the air speed.

### 2.3.3 Operative temperature

The operative temperature ($T_{op}$) is a measure that combines the air temperature and the mean radiant temperature into a single value to express their joint effect. It is a weighted average of the two, the weights depending on the heat transfer coefficients by convection ($h_c$) and by radiation ($h_r$) at the clothed surface of the occupant. It is often used to express the temperature of a space. Thus in both the ASHRAE 55 and the CEN EN15251 standards the comfort temperature in their adaptive sections is expressed in terms of the operative temperature (see Figures 5.1 and 5.2).

### 2.3.4 Evaporation

When water evaporates it extracts a quantity of heat (the *latent heat of evaporation*) from its surroundings. Evaporation is an endothermic process – called so because energy, in the form of heat, is required to change water into a vapour. This cooling effect is very powerful, and cools the body and the air in its vicinity when we sweat. It is not the sweating that cools us but the evaporation of the sweat from the skin. Evaporative cooling becomes increasingly important as ambient operative temperatures rise through and above skin temperatures, from around 28°C through to 35°C, above which temperature the body relies solely on evaporation to cool itself.

Once the sweat has evaporated it must move away from the skin in order that more evaporation can occur. In still air this is done by the natural convection mentioned in Section 2.3.2. Evaporation is driven by differences in water vapour pressure. The water vapour pressure is that part of the total pressure of the air that is caused by the molecules of water in it. The hotter the air, the more water it is capable of carrying. Air next to the skin is close to skin temperature so the maximum water vapour pressure it can hold (the 'saturated vapour pressure') is determined largely by skin temperature. The driving force for the evaporation of sweat is the difference between this saturated vapour pressure at skin temperature and the vapour pressure in the air as a whole.

It is as well to remember that the total heat lost by evaporation is determined by the amount of sweat actually evaporated and not by the maximum that could be evaporated (this would require the entire body surface to be wet with sweat). So the sweating is adjusted by the thermoregulatory system to that amount needed for the body to maintain its thermal equilibrium.

A factor *skin wettedness* (w) has been devised which relates the actual sweating heat loss to the maximum possible. It (w) has been used as a measure of heat stress because it is an indication of how near the person is to the limit of the rate at which heat can be lost from the skin through evaporation. This maximum rate depends on factors including the air speed and the nature of the clothing. As people become acclimatised to higher temperatures, they may begin to sweat at a lower body-core temperature. Their rates of sweating change as does the distribution of sweat production areas on the body, as sweat glands are 'trained' to maximise their cooling effect, and so reduce the body's heat stress.

Social attitudes to the skin wetness caused by sweat vary. In temperate climates sweaty skin is considered uncomfortable, while in warm humid climates wet skin can be considered the norm, and a dry skin can be considered uncomfortable.

Air movement tends to increase evaporation rates in much the same way as it increases convective heat loss. Possible rates of heat loss by evaporation are driven by the water vapour pressure in the atmosphere and the square root of the air velocity. Skin wettedness is a factor in commonly used heat-transfer equations, but it will only affect the thermal balance when its value approaches unity, implying that the body is 'having trouble' evaporating all the sweat it is producing.

Air movement can be particularly pleasant when the air temperature is a little lower than the mean skin temperatures, which range from around 32°C to 35°C. At these warm temperatures comfort can be greatly enhanced by sitting in a breeze. The faster the air moves over the skin, the faster heat is lost from the body, by evaporation as well as by convection, so keeping it cool. These are the conditions that the ancient Greeks would have written about in their poems in praise of the delightful 'Zephyrs' or cooling breezes.

### 2.3.5 Insensible and respiratory heat loss

There is also some heat loss by evaporation of moisture that diffuses through the skin even when there is no sweating. The process is called insensible perspiration.

There is also loss of heat and moisture from the lungs and upper respiratory tract during breathing. The air breathed in is warmed in the lungs to near the core temperature and moistened to almost complete saturation. Because the rate of breathing depends on metabolic rate, the rate of heat loss from the lungs depends on the metabolic rate. The evaporative heat lost is proportional to the difference between the saturated water vapour pressure at core temperature and the vapour pressure in the surrounding air. The respiratory contribution to heat loss can be a substantial part of the whole, particularly in dry conditions, and is some 20–30 per cent for a sedentary person in thermal comfort.

## 2.3.6 Clothing

Outside the tropics, where the air is typically close to body temperature, clothing plays a major role in enabling us to survive. As we shall see in Chapter 3, it is of particular importance to the adaptive model of thermal comfort. In calculating heat loss from the body surface, clothing is usually treated as if it were a uniform layer of insulation between the body and the environment, and having a single surface temperature ($T_{cl}$). Quite clearly this is an approximate treatment. Clothing is actually anything but uniform. The face and hands are generally uncovered and the actual clothing insulation varies from place to place on the body according to the nature of the clothing worn. In practice, however, the assumption works quite well, and the overall insulation of the clothing can be expressed as the sum of the contributions from the individual items of clothing being worn, as if they were each spread over the whole of the body surface (Lotens, 1989). The layers of air trapped between multiple layers of clothing and between the clothing and the skin are counted as part of the clothing ensemble.

In thermal comfort studies the insulation of the clothing is generally expressed in clo units. The clo unit was introduced to facilitate the visualisation of clothing insulation, and is the insulation necessary to keep a seated person comfortable at 21°C – about that of an office worker's suit with a shirt and the usual underwear. A thermal insulation of 1 clo = 0.155 m².K/W. Tables of clothing insulation values for typical western ensembles are given in Table 8.1 and the range of local clothing types is continually being added to (Figure 2.4). One problem with such tables is that the meaning of the clothing description can vary according to culture and climate, so a suit described as thick in one climate might be counted rather thin in another.

In addition to acting as insulation against the transfer of dry heat, clothing can affect heat loss by evaporation. First, the clothing introduces extra resistance to the diffusion of water vapour away from the skin. The strength of this effect depends on the design of the clothing and the permeability to moisture of the material it is made of. Second, clothing can absorb excess moisture next to the skin. The absorbed moisture is then evaporated from the clothing and not from the skin, so the latent heat loss pathway is less direct.

The actual way clothing works is often far more complex than the classic model outlined above indicates. In some hot dry climates, for instance, the inhabitants wear loose, multiple-layered clothing. The function of the clothing is then to keep the high environmental temperatures and solar radiation away from the skin, whilst still allowing heat loss by evaporation as air is pumped over the body surface and through the clothing when the wearer moves (Berger, 1988). Where there is a high rate of sweating the heat loss can actually be increased by the clothing, because it provides extra surfaces from which evaporation can take place, so further cooling the space between the skin and the clothing.

| Ensemble type | Ensemble description | $f_{cl}$ | $I_a$ | | $I_{cl}$ | | $I_T$ | |
|---|---|---|---|---|---|---|---|---|
| | | | clo | m²·K·W⁻¹ | clo | m²·K·W⁻¹ | clo | m²·K·W⁻¹ |
| Male summer clothing | T-shirt with 1/3 sleeves, short serwal, thowb and sandals | 1,30 | 0,594 | 0,092 | 0,59 | 0,092 | 1,05 | 0,163 |
| | T-shirt with 1/3 sleeves, short serwal, thowb, kuffiya, white ghutra, eqal, sandals | 1,35 | 0,594 | 0,092 | 0,69 | 0,107 | 1,13 | 0,175 |
| | T-shirt with 1/3 sleeves, short serwal, long serwal, thowb, kuffiya, white ghutra, eqal, sandals | 1,36 | 0,594 | 0,092 | 0,79 | 0,123 | 1,23 | 0,191 |
| Male winter clothing | T-shirt with 1/3 sleeves, short serwal, long cotton serwal, thowb, kuffiya, ghutra shemagh, eqal, shoes | 1,46 | 0,594 | 0,092 | 0,84 | 0,131 | 1,25 | 0,194 |
| | T-shirt with 1/3 sleeves, short serwal, long cotton serwal, thowb, ghutra shemagh, eqal, jacket, shoes | 1,45 | 0,594 | 0,092 | 1,29 | 0,200 | 1,70 | 0,264 |

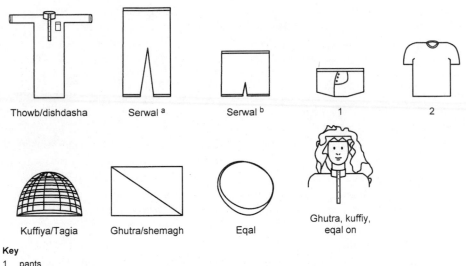

Thowb/dishdasha    Serwal [a]    Serwal [b]    1    2

Kuffiya/Tagia    Ghutra/shemagh    Eqal    Ghutra, kuffiy, eqal on

**Key**
1   pants
2   T-shirt
[a]   Long
[b]   Short

**FIGURE 2.4** Clothing insulation levels change according to culture as well as climate

Source: ISO 9920.

The need to allow for clothing insulation and permeability is a source of considerable uncertainty in applying the physical model of human heat exchange. An upholstered chair can also contribute as much as 0.4 clo to the 'clothing' insulation (depending on the type of chair) and the appropriate extra insulation should be included when estimating clothing insulation of seated people.

Another complication in dealing with clothing is that its function is not purely thermal. We wear clothes for social as well as thermal purposes. Small variations such as wearing a jacket open or closed makes a significant difference to its insulation as can the wearing, or not wearing, of a tie. In most cultures there is a minimum of clothing that is socially acceptable. Things done for social reasons have physical effects.

## 2.4 Behaviour

Before we introduce the various approaches to thermal comfort studies in Chapters 3 and 4 it is worth noting that behaviour also plays an important role in our thermal interaction with the environment. Ken Parsons (2003) suggests that 'A most powerful form of human thermoregulation is behavioural – putting on or taking off clothing, change posture, move, take shelter etc.'. All the approaches to our thermal interactions with the environment listed above have in effect assumed that we are acted upon by the environment and react to it in a passive way. In fact we have a very active interaction with our environment.

We deal with such behavioural actions more fully in the next chapter but in essence thermal interactions take a number of forms:

- Clothing changes
- Changes of posture and metabolic rate
- Movement between different thermal environments
- Making use of thermal controls to change the current environment.

All of these are deliberate actions to control the environment and augment the unconscious physiological reactions already referred to.

Time is important for behavioural interactions. There are four typical time periods for these effects:

- Immediate – the change of clothing in anticipation of a thermal change; for example, putting on a coat before going out.
- Within-day – the clothing changes, changes of posture or environmental adjustments we use to cope with changing environments within a particular day.
- Day-to-day – we learn from one day to the next how to cope with changing conditions such as the weather.
- Longer term – seasonal changes in clothing, in the use of buildings, activities learned over a longer period.

In order to fully describe people's thermal experience we need to take account of all these changes. The total picture must be consistent with the findings drawn from physics and physiology, but will change with climate, place and time in a dynamic and interactive way. These considerations are addressed in greater detail in Chapter 4.

## 2.5 Equations for heat balance in the human body

Different sources do not always agree on the exact formulation of the human heat balance equation. We have used information from Parsons (2003) and the ASHRAE Handbook of Fundamentals (2009).

The basic equation for thermal balance is:

$$M - W = C + R + E + (C_{res} + E_{res}) + S \qquad (2.1)$$

where  M  is the metabolic rate

       W  is  mechanical work done

       C  is  convective heat loss from the clothed body

       R  is  radiative heat loss from the clothed body

       E  is  evaporative heat loss from the clothed body (sweat and insensible evaporation)

       $C_{res}$  is convective heat loss from respiration

       $E_{res}$  is evaporative heat loss from respiration

       S  is  the rate at which heat is stored in the body tissues.

The equations for radiative heat transfer are

$$R = \varepsilon h_r f_{cl} f_{eff} (T_{cl} - T_r) \ W/m^2 \qquad (2.2)$$

where  R  = rate of radiative heat flow per square metre of body surface

      $\varepsilon$  = the emissivity of the clothed/skin surface (generally close to 1)

      $h_r$  = the linear radiation transfer coefficient ($W/m^2K$)

      $f_{cl}$  = the effective surface area of the clothed body (greater than one because the surface area of a clothed body is usually greater than a nude one)

      $f_{eff}$  = the effective radiation area factor (less than one because in some parts – e.g. under the arms – the body radiates to itself rather than to the environment)

      $T_{cl}$  = the clothing surface temperature (°C)

      $T_r$  = the mean radiant temperature (°C ).

As a first approximation:

$$f_{cl} = 1 + 0.15 \ I_{clo} \qquad (2.3)$$

where  $I_{clo}$  = the clothing insulation (see Chapter 8 for tables)

$$f_{eff} = 0.7 \text{ when seated, } 0.72 \text{ when standing (Fanger, 1970)} \qquad (2.4)$$
$$h_r = 4.7 \ W/m^2K \text{ for typical clothing} \qquad (2.5)$$

The equation for convective heat transfer is

$$C = h_c (T_{cl} - T_a) \ W/m^2 \qquad (2.6)$$

where  C  is the rate of convective heat flow per square metre of body surface

      $T_{cl}$  = the mean surface temperature of the clothed body

      $T_a$  = the air temperature

      $h_c$  = the convective transfer coefficient in $W/m^2K$.

$h_c$ is itself dependent on $T_{cl} - T_a$, and the air velocity, but the dependence on temperature is small so that $h_c$ is generally taken to be given by:

$$h_c \approx 3.1 \ W/m^2K \ for \ v < 0.2m/s \ (value \ for \ standing \ person) \qquad (2.7)$$

$$h_c = 8.3 \ v^{0.6} \ W/m^2K \ for \ v > 0.2m/s \qquad (2.8)$$

The constant value at low air speeds reflects air currents from natural convection. The operative temperature is defined as:

$$T_{op} = H \ T_a + (1 - H) \ T_r \qquad (2.9)$$

where $H$ = the ratio $h_c/(h_c + h_r)$ .

Researchers have differed in their estimates of the values of these heat transfer coefficients, and hence of the value of H. In the CIBSE Guide A (CIBSE, 2006) the value $\sqrt{(10 \ v)}$ is used for the ratio of $h_c$ to $h_r$ where v is the air speed, and so:

$$T_{op} = \frac{T_a \ \sqrt{(10 \ v)} + T_r}{1 + \sqrt{(10 \ v)}} \qquad (2.10)$$

At indoor air speeds below 0.1m/s, natural convection is assumed to be equivalent to v = 0.1, and equation 2.8 becomes:

$$T_{op} = \tfrac{1}{2} \ T_a + \tfrac{1}{2} \ T_r \qquad (2.11)$$

Operative temperature is a theoretical and not an empirical measure and therefore cannot strictly be measured directly but in practice it approximates closely to the globe temperature (see above). In well-insulated buildings and away from direct radiation from the sun or from other high temperature radiant sources, the difference between the air and the mean radiant temperatures (and hence between the air and operative temperatures) is small.

The equations for evaporative heat loss are

$$E = wh_e(p_{ssk} - p_a) \qquad (2.12)$$

where  E   is  the rate of evaporative heat flow per square metre of body surface
       w   is  the skin wettedness
       $h_e$  is  the evaporative heat transfer coefficient (see below)
       $P_{ssk}$ is  the saturated water vapour pressure at skin temperature ($kP_a$)
       $P_a$  is  the water vapour pressure of the air ($kP_a$).

$$h_e = 16.5h_c \ W/m^2kP_a \qquad (2.13)$$

The equations for insensible loss have been determined empirically as:

$$E_{is} = 4 + 0.12 \ (p_{ssk} - p_a) \ W/m^2 \qquad (2.14)$$

The respiratory heat loss has two components, latent ($E_{res}$) owing to evaporation and dry ($C_{res}$) owing to convective heat transfer in the lungs:

$$E_{res} = 0.017M(5.87 - p_a) \ W/m^2 \qquad (2.15)$$

$$C_{res} = 0.0014M(34 - T_a) \ W/m^2 \tag{2.16}$$

M is the metabolic rate in $W/m^2$ (the rate of heat loss is affected by respiration and thus by metabolic rate).

The equations for heat transfer across clothing are:

$$K = R + C = (T_{sk} - T_{cl})/(0.155I_{clo}) \ W/m^2 \tag{2.17}$$

where K = the conducted heat through the clothing (equal to the sum of the radiative (R) and convective (C) losses from the clothing surface)

$I_{clo}$ = the clothing insulation in clo units.

The maximum rate of evaporation is given by

$$E_{max} = f_{pcl}h_c(p_{ssk} - p_a) \ W/m^2 \tag{2.18}$$

where $E_{max}$ = the maximum evaporation from the clothed body surface (below this value the evaporation is physiologically and not physically determined)

$f_{pcl}$ = the permeability efficiency factor of the clothing ensemble (= $1/(1 + 0.143h_cI_{clo})$).

The equation for estimating the DuBois surface area of the human body (Dubois and Dubois, 1916) is

$$A_d = 0.202wt^{0.425}l^{0.725} \ (m^2) \tag{2.19}$$

where wt = the weight in Kg and l is the height in metres.

## REFERENCES

ASHRAE (2009) *ASHRAE Handbook of Fundamentals.* Atlanta, Georgia: American Society of Refrigerating and Air Conditioning Engineers.

Auliciems, A. and Szokolay, S.V. (2007) *Thermal Comfort* (2nd edn). PLEA Passive and Low Energy Architecture International. Available at www.plea-arch.org.

Berger, X. (1988) The pumping effect of clothing: *Int J Ambient Energy*, 9(1), 37–46.

Cabanac, M. (1992) What is sensation? In R. Wong (ed.) *Biological Perspectives on Motivated Activities.* Northwood, NJ: Ablex.

Cândido, C., de Dear, R.J., Lamberts, R. and Bittencourt, C. (2010) Cooling exposure in hot humid climates: Are occupants 'addicted'? In S. Roaf (ed.) *Transforming Markets in the Built Environment: Adapting for climate change.* London: Earthscan, ch. 6.

CIBSE (2006) *Environmental criteria for design. Chapter 1: Environmental Design: CIBSE Guide A.* London: Chartered Institution of Building Services Engineers.

Clark, R.J. and Edholm, O.G. (1985) *Man and his Thermal Environment.* London: Arnold.

DuBois, D. and DuBois, E.F. (1916) A formula to estimate surface area if height and weight are known, *Archives of Internal Medicine* 17, 863.

Fanger, P.O. (1970) *Thermal Comfort.* Copenhagen: Danish Technical Press.

Hopkinson, R.G. (1963) *Architectural Physics: Lighting.* London: HMSO.

Lotens, W.A. (1989) The actual insulation of multilayer clothing, *Scand J Work Environ Health* 15(1), 66–75.

Miller, G. (1956) The magic number seven, plus or minus 2, *Psychological Review* 67, 81–97.

Parsons, K. (2003) *Human Thermal Environments.* Oxford: Blackwell.

# 3

# FIELD STUDIES AND THE ADAPTIVE APPROACH

If we want to know how people feel in a particular situation there is no better way to find out than to go and ask them. This is the method of the field survey. It is the basic tool of the adaptive approach (Humphreys, 1995). This is why the second half of this volume is a guide to how to conduct, analyse, interpret and report on field surveys. An adaptive approach to comfort is essential; it is impossible, using only a simple theoretical model (see Chapter 4), to understand the complex workings of a local comfort system that involves ever-changing people, buildings and climates. So a method is needed that enables us to record and analyse local conditions and behaviours, and so provide an effective model to help design for site-specific comfort. In this book we hope to meet the need for a clear exposition of an adaptive model of thermal comfort, useful alike for research and practical application. Because context, culture, buildings and climate are unique to any particular place, so also are the comfort needs and expectations of its inhabitants. For this reason we would encourage teachers, researchers, students, architects, engineers and others to undertake their own comfort research in their own towns and regions to see the complexities of the comfort system they themselves inhabit.

## 3.1 Field surveys of thermal comfort

A field survey of thermal comfort is an *in situ* poll of comfort (Table 2.1) among a given population (say the workers in an office) together with simultaneous measurements of the environmental conditions. In field surveys participants wear their normal clothing and go about their usual work. There has been a continuing stream of thermal comfort field surveys reaching right back to the early decades of the last century. The results of such surveys are analysed statistically to estimate the temperature[1] at which the average survey participant will be comfortable, usually called the 'comfort temperature' or the 'neutral temperature' – or, alternatively, the temperature at which the largest number of participants will be comfortable.

---

1  We use 'temperature' as shorthand for the value of a thermal index that incorporates one or more of the four principal components of the thermal environment: air temperature, mean radiant temperature, air movement and humidity.

Alongside the field studies are the laboratory-based heat balance models of thermal comfort described in Chapter 4.

Field surveys have been used to develop empirical, statistically based indices of thermal comfort. Bedford's Equivalent Temperature (Bedford, 1936) and Sharma and Ali's Tropical Summer Index (Sharma and Ali, 1986) are examples. Such indices share the objective of the theoretical heat-balance indices to estimate the conditions which people find comfortable. However, they suffer the limitation that they apply only to the conditions and context of the field survey on which they are based and it is virtually impossible to accurately extrapolate those results to fit the conditions that exist in another location and site.

A survey questionnaire can be extended to include, say, the sex, age and weight of the participants or the time of day at which they are polled, so that the effect of these factors on the resulting comfort temperature can be evaluated. Studies have been done to explore the effects of over 60 factors, including personality type, management structures and the colour of the walls, but these factors are not central to most surveys. Many recent survey questionnaires have also included the ways in which the participants interact with the buildings they occupy, such as when they are in the building, when they open and close windows, turn the heating on or off, shade the windows and so on. A background descriptive survey may also be made on the building being occupied – is it light- or heavyweight? Does it use central air conditioning? Or natural ventilation? And so on. A thermal comfort survey can be part of a more comprehensive survey of the environment that includes, for example, noise, lighting and air quality.

Surveys are of two different basic patterns – longitudinal surveys where a relatively small number of subjects are polled for their comfort vote repeatedly over an extended period of time, and transverse surveys where a large number of people are polled just once within a limited survey period. Some projects have used both kinds of survey. A discussion of the relative merits of the two patterns is given in Chapter 8.

The second part of this book is a simple introduction to field survey methodology and analysis. The thermal comfort survey has at its heart the comfort of the people being surveyed. In addition to being a tool for investigating thermal comfort, a field survey can tell us a lot about the way in which a building works to provide comfort for its occupants, or maybe works against them. It can also tell much about the best way to design buildings that are appropriate to the different climates and cultures of peoples in different parts of the world. The second volume of this trilogy on comfort presents more detailed methods to use in the analysis and interpretation of results from field surveys. Volume three builds on Chapter 6 to explore the lessons we can learn from field surveys about how to design better buildings and why local buildings may succeed or fail in providing comfort.

## 3.2 Post-occupancy surveys

Another type of survey, the Post-occupancy Evaluation (POE) is essentially concerned with the individual building in which the survey is being conducted (Nicol and Roaf, 2005). It is also called Building Performance Evaluation by some practitioners (Preiser and Vischer, 2004). In particular, the evaluation may be concerned with energy use and the effectiveness of any energy-saving technologies, as well as whole building performance. The normal post-occupancy questionnaire will ask building occupants about their experience of the building – whether it gets too hot in summer or too cold in winter, is too noisy or too bright, for instance – and the results can be used to assess the relative success of the design in relation to its overall environmental performance. Post-occupancy studies have identified that occupants are more 'forgiving' of, and work more efficiently in, buildings they like. This may be the result of a popular management system, or a very quiet, peaceful building, or very often one that affords them more control over their own environments (Leaman and Bordass, 1997). These studies highlight the importance of occupant behaviours and perceptions, and well as the performance of the building, on the overall comfort experience of people in buildings. The best source of information on Post-occupancy Evaluation is the Usable Buildings Trust (2011).

## 3.3 Comfort and indoor temperature: the basic adaptive relationship

Figure 3.1 illustrates the basic relationship which has been shown to arise in meta-analyses of field study results. Note that in this and in all the scatter graphs in this chapter each point represents the results for a whole survey which will contain numerous comfort votes and associated environmental measurements. Each value for comfort temperature is based on no fewer than 20 and maybe as many as hundreds of votes and the error is estimated to be 0.4K at most (see Figure 9.1). The central line shows the mean neutral temperature for any given mean operative temperature and 95 per cent of all the results lie between the two outer lines. The figure shows that the comfort or neutral temperature calculated from a field survey is highly correlated with, and often almost equal to, the mean operative temperature ($T_{op}$ – see Section 2.3.3) measured

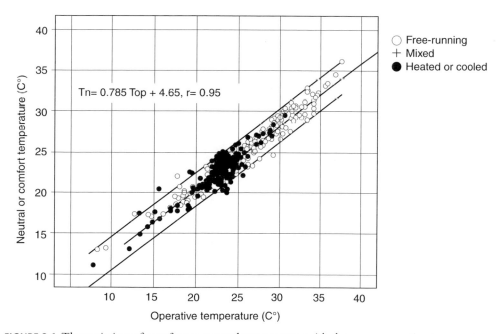

**FIGURE 3.1** The variation of comfort or neutral temperature with the mean operative temperature ($T_{op}$) in a large number of survey populations

Source: M. Humphreys.

during the survey. Over time people are usually able to match their comfort temperature to their normal environment. This graph distinguishes between surveys according to whether the buildings they were conducted in were in the free-running mode (no heating or cooling at the time of the survey), heated/cooled mode or some mixture of the two. It is noticeable that the adaptive effect works equally in free-running, heated or cooled, or mixed mode.

Outside a range of mean operative temperatures of 20°C to 25°C, the match of comfort temperature to mean operative is not exact. In hotter environments the comfort temperature is generally lower than the mean operative temperature, suggesting that the people would be generally slightly warmer than they would like to be and in colder environments they are generally slightly cooler. Note though that the values of neutral temperature in this analysis were calculated for still air. In fact, most offices and homes in hot climates have fans available which will raise the comfort temperatures by about 2K (Nicol, 2004) bringing the comfort temperature and the mean operative temperature even closer together.

With the need to reduce energy consumption in buildings, research has increasingly turned to field studies to establish appropriate comfort conditions for local populations. Setting the results of those field studies into a universal adaptive model is more problematic. This is partly because collecting together all the data is difficult. Field study research has been undertaken by many workers from different disciplines working in diverse places, and their results, together with the various comfort models they have used, are scattered among journals relating to ergonomics, physiology, climatology, engineering, building science and architecture.

## 3.4 Outcomes: indoor comfort and outdoor temperature

Once the relationship between mean indoor temperature and comfort temperature had been established (Humphreys, 1976) it was found that there was also a strong relationship between indoor comfort temperature and the outdoor temperature. This allows the effect of climate on comfort temperature to be estimated. Humphreys (1978) collected data from reports of field surveys from all over the world and produced the well-known graph in Figure 3.2. Note that the comfort temperature in free-running buildings varies linearly with the outdoor temperature whilst that of heated buildings shows a more complex curvilinear relationship. Few data were available at that time for comfort in cooled buildings in hot climates and it was assumed that they would continue the trend of the heated or cooled buildings.

The relationship shown in Figure 3.2 can explain the difference observed between comfort temperatures in buildings in the heated and in the free-running mode. The indoor temperature in free-running buildings is coupled to the outdoor temperature through the fabric of the building whereas that of a building in heated (or cooled) mode is decoupled by the closed skin of the building and the operation of the heating and cooling systems within it. People indoors in a free-running building are adapting to the external temperature around the building mediated through its walls and openable windows, and through roofs and floors. The temperature in a heated or cooled building is controlled by the thermostat, often set by the building manager, and its occupants adapt to this pre-selected temperature.

In Figure 3.2 the range of comfort temperatures at any one outdoor temperature is real, and not just a scatter of 'errors' about a mean. It suggests that there is a range (of about 4K) of mean temperatures people find comfortable. A recent paper by Humphreys et al. (2010), using results from comfort surveys collected since 1978, has confirmed the general shape of

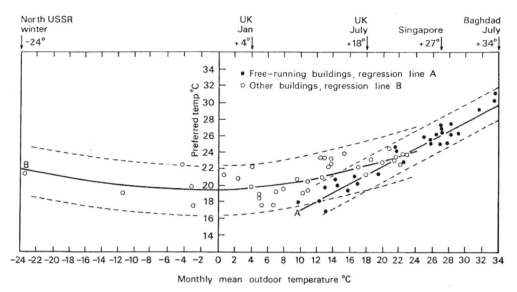

FIGURE 3.2 Humphreys' graph of 1978 showing how the mean indoor comfort temperature varies with monthly mean outdoor temperature in buildings that are in free-running mode (filled dots) and heated or cooled mode (open dots). The continuous lines are the average of the values and 95 per cent of values lie within the dashed lines

Source: HMSO.

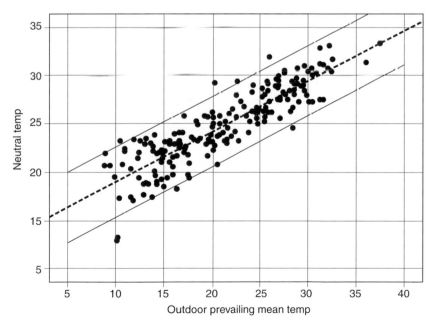

FIGURE 3.3 Scatter of neutral temperatures and the prevailing mean outdoor temperatures in buildings in free-running mode. Data collected from field surveys since 1978

Source: after Humphreys et al., 2010.

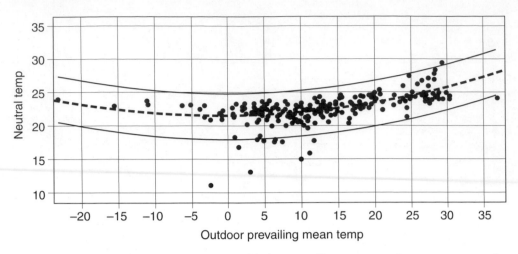

FIGURE 3.4 Scatter of neutral temperatures and the prevailing mean outdoor temperatures in buildings in heated or cooled mode since 1978

Source: Humphreys *et al.*, 2010.

the graph in Figure 3.2. However, Figures 3.3 and 3.4 from that paper show that the mean comfort temperature for any given outdoor temperature has risen by about 2K in buildings during that period in both free-running and in heated and cooled modes. This is possibly because buildings have become warmer, and people have adapted to these higher indoor temperatures. It also suggests that more recent building types may provide less protection against summer heat and that, in general, free-running buildings are running hotter because of this. However, too much should not be made of this increase, since the surveys in the meta-analyses on which this finding was based were done in buildings that were not selected on a strictly representative basis.

The range of comfort temperatures at any given outdoor temperature in heated and cooled buildings remains much the same as in Figure 3.2 but in free-running buildings it has increased from about 4K in pre-1978 buildings to about 7K (Figure 3.3). This would be expected if building thermal designs were more diverse than previously.

The strong relationship between indoor comfort and outdoor temperature has been the basis of adaptive standards for indoor temperatures and this aspect is covered in more detail in Chapter 5.

## 3.5 The basis of the adaptive model: using surveys to understand comfort

We start with the biological insight that the human being is a comfort-seeking animal who will, given the opportunity, interact with the environment in ways that secure comfort (Levins and Lewontin, 1985). Nicol and Humphreys (1973) suggested a feedback approach to interpreting the results of field surveys of thermal comfort. The approach viewed sensations of heat or cold as an active part of a comfort control system. Unpleasant sensations prompt reactions from people and cause them to make changes in the comfort control system itself. In addition to the automatic physiological response of the body, people's conscious behavioural actions will alter their relationship to the world around them, ultimately helping to safeguard the core temperature of the body. The greater the difference between the core temperature and the environmental temperature, the greater the task of the comfort control system.

This way of interpreting thermal comfort has become generally known as the Adaptive Model and is governed by the adaptive principle:

> *If a change occurs such as to produce discomfort, people react in ways which tend to restore their comfort.*

Humphreys and Nicol (1998) give the following lists of the actions that might be initiated in response to cold or to heat. The lists contain physiological, psychological, social and behavioural items. They are based on the climate and culture of the UK and some may need modification in other places.

*Some conceivable actions in response to cold:*

- Vasoconstriction (reduces blood flow to the surface tissues)
- Increasing muscle tension and shivering (generates more heat in the muscles)
- Curling up or cuddling up (reducing the surface area available for heat loss)
- Increasing the level of activity (generates body heat)
- Adding clothing (reduces the rate of heat loss per unit area)
- Turning up the thermostat or lighting a fire (usually raises the room temperature)
- Finding a warmer spot in the house or going to bed (select a warmer environment)
- Visiting a friend or going to the library (hoping for a warmer environment)
- Complaining to the management (hoping someone else will raise the temperature)
- Insulating the loft or the wall cavities (hoping to raise the indoor temperature)
- Improving the windows and doors (to raise temperatures/reduce draughts)
- Building a new house (planning to have a warmer room temperature)
- Emigrating (seeking a warmer place long term)
- Acclimatising (letting body and mind become more resistant to cold stress).

*Some conceivable actions in response to heat:*

- Vasodilation (increases blood flow to surface tissues)
- Sweating (evaporative cooling)
- Adopting an open posture (increases the area available for heat loss)
- Taking off some clothing (increases heat loss)
- Reducing the level of activity (reduces bodily heat production)
- Having a beer (induces sweating and increases heat loss)
- Drinking a cup of tea (induces sweating, more than compensating for its heat)
- Eating less (reduces body heat production)
- Adopting the siesta routine (matches the activity to the thermal environment)
- Turning on the air conditioner (lowers the air temperature)
- Switching on a fan (increases air movement, increasing heat loss)
- Opening a window (reduces indoor temperature and increases breeze)
- Finding a cool spot or visiting a friend (hoping for a cooler temperature)
- Going for a swim (selects a cooler environment)

- Building a better building (long-term way of finding a cooler spot)
- Emigrating (long-term way of finding a cooler place)
- Acclimatising (letting body and mind adjust so that heat is less stressful).

There are five basic types of adaptive actions:

1 Regulating the rate of internal heat generation
2 Regulating the rate of body heat loss
3 Regulating the thermal environment
4 Selecting a different thermal environment
5 Modifying the body's physiological comfort conditions.

Some actions such as vasodilation and vasoconstriction, sweating and shivering, are not under conscious control. Changes of posture or physical activity may be either deliberate or unconscious. Even some of the overtly behavioural responses may become 'second nature' in a particular society, culture or climate.

There are so many possible actions that comfort will often be restored by a set of minor actions rather than by a single one. For example, a response to cold might be a slight increase in muscle tension, a barely perceptible vasoconstriction, a slightly 'tighter' posture, putting on a sweater and the desire for a cup of coffee. Each might be small, but the joint effect can be large, especially as changes in heat flow and thermal insulation are multiplicative in their effect (see Figure 3.5 in Plates). Questionnaire responses about clothing and activity would not pick up the more subtle adaptations, and they would be 'invisible' to the procedures for the evaluation of the steady-state physiological models (see Chapter 4).

Adaptation can be regarded as a set of learning processes and people become well adapted to their usual environments. They will feel hot when the environment is hotter than 'usual' and cold when it is colder than 'usual'. The adaptive approach is therefore interested in the study of 'usual' environments. In particular, what environments are 'usual'? How does an environment become 'usual' and how does a person move from one 'usual' environment to another? We explore the implications of this insight for ensuring comfortable buildings in Section 3.8.

Auliciems (1981) and de Dear and Brager (1998) are among others who took up the approach. They have generally put a greater emphasis on the role of psychophysics and expectation than Humphreys and Nicol who tend to attribute adaptation to the small, often overlooked or unmeasured, physical actions people take to remain comfortable. Recent comfort research has put increasing emphasis on the adaptive approach and its key operating mechanisms – the behaviours people use to ensure their own comfort.

## 3.6 Opening the black box

One criticism of the adaptive comfort model is that it is a 'black box' based on empirical observations in which the nature of the adaptive mechanisms is hidden, and not quantified or related to measurements. One answer to this criticism is to develop models of the operating patterns of individual contributory adaptation mechanisms, such as opening windows, pulling down the blinds or turning on the heating, that are clearly a part of the whole adaptive comfort system. Examples of such models are those of Nicol and Humphreys (2004), Yun and Steemers (2007), Haldi and Robinson (2010) and the 'Humphreys' algorithm developed in Rijal *et al.* (2012). Considerable research is now being done to incorporate such models, mostly for window-opening behaviours, into building simulation programs. The Humphreys algorithm model is available in the esp-r simulation tool (Clarke, 2001). Many such algorithms that integrate adaptive behaviour into building simulation models are in development and will increasingly be the subject of research and discussion.

Humphreys and his team have developed an algorithm (Rijal *et al.*, 2012) to predict the use of windows and other low-energy adaptive actions in occupied buildings. The algorithm is based on the adaptive principle assuming that people open windows in order to avoid discomfort indoors. Figure 3.6 shows the basic reasoning behind the model. Figure 3.7 shows how the basic 'hysteresis' shape predicted by Humphreys is borne out by data from the field. The model enables computer simulations of rooms with windows to estimate the likelihood that a window will be open and to make the appropriate adjustment to the energy flows and the indoor temperature. Among others Yun and Steemers (2007) have used a method based on probability density functions derived from surveys using Markov chains to develop a model on which the probability of windows being open is developed. The basic form of the

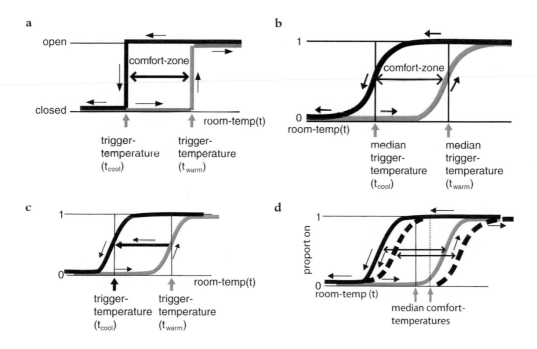

**FIGURE 3.6** Four steps in the development of the Humphreys algorithm: a) an individual will open a window as the indoor temperature becomes too hot (or close it as it becomes cold); b) several individuals will become more likely to open the window as more of them feel too hot, giving a hysteresis shape which will contain all possibilities; c) because not all windows are necessarily open if the temperature does not rise too high the actual number of windows open can fall anywhere within the hysteresis 'loop'; d) a constraint on window-opening will mean that occupants will not open their window so soon and will shift the loop to a higher temperature

Source: H. Rijal.

Longitudinal UK data: indoor globe temperature and proportion of windows open

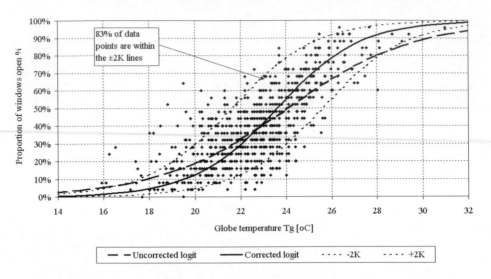

**FIGURE 3.7** Data from UK offices with opening windows confirms the basic shape of the Humphreys algorithm. The proportions of windows open are calculated and plotted against indoor globe (see Chapter 8) temperature

Source: H. Rijal.

models for window-opening behaviour are becoming clear, but there is room for them to be adjusted in the light of experience or further information.

Another adaptive action that has already been widely investigated is the use of clothes. The use of clothing among UK schoolchildren was the subject of surveys by Humphreys (1973) and his co-workers. Morgan *et al.* (2002) are among others who have modelled and analysed the use of clothing through field surveys. Changes of posture are among the reactions to discomfort and Raja and Nicol (1997) looked at ways in which posture could be recorded and related to temperature.

A further problem for a model of behaviours, especially when there are a number of ways to achieve a particular goal (e.g. to switch on a fan or to open a window), is to decide how to allow for the different possibilities and the likelihood that people will choose one option over another. One approach is to assess the constraints that act upon the person to prevent or discourage a particular action (Rijal *et al.*, 2012). There is great scope for development of the adaptive model and to clarify and quantify the adaptive processes. A better understanding of these processes will greatly assist the design and operation of comfortable low-energy buildings.

### 3.6.1 Extending the applicability

Another issue that has been raised about the adaptive model is that it overlooks the effect of humidity and air movement on the temperature at which people are comfortable. It need not do so. In principle there is no difficulty in including these effects. Nicol (2004) has looked at the effect of these factors on comfort in hot climates and has suggested that air movement created by fans increases the comfort temperature by an average of about 2K, but the effect of high humidity is less consistent. There is a persistent perception that high humidity makes us feel hotter, and there may be other factors, such as sweatiness and nasal thermal perception, that are being experienced. In ordinary environments the effect of humidity on our thermal perception as measured by comfort vote is small, and high humidity may reduce the range of comfort temperature rather than shifting it up or down (de Dear et al., 1991). Other researchers have also questioned the importance of humidity in the perception of comfort (Givoni and Belding, 1962).

Givoni comments:

> In this research I have found, experimentally, that humidity does not affect the rate of evaporation from the body, except under extreme conditions, because the body has the capability to compensate for higher humidity by increasing its wetted area. A moist and even wet skin is thus of physiological benefit, but we don't like it. Consequently, the effect of humidity is mainly a psychological one, as we don't like the feeling of a moist skin. In the several comfort studies in hot countries in which I participated, in Thailand (Givoni et al., 2004) and Hong Kong (Givoni et al., 2005), it was demonstrated that people acclimatized to a humid climate are much less sensitive to the effect of humidity in comparison with the assumptions in various comfort indices.
>
> (Personal communication, 2011)

These observations accord with our discussion of evaporative heat loss in Section 2.3.4. However, it should be noted that at air temperatures above the body core temperature, very high humidities are lethal.

## 3.7 Adaptive comfort and non-standard buildings

Whatever the type of building, adaptation itself is not a problem for the inhabitants so long as there is a sufficient range of appropriate 'adaptive opportunity' (see glossary) (Baker and Standeven, 1995) to allow inhabitants to adapt to indoor and outdoor temperature changes over time. There are obviously problems when spikes of temperature are experienced, as in heatwaves, and also if the switch from one mode to another is not properly handled. It is well known that it is the 'shoulder months' of spring and autumn, during which the outdoor temperature changes quite rapidly, that are often the problem for comfort. As we show in the next chapter, comfort standards treat the two modes differently.

### 3.7.1 Passively cooled or heated buildings

A 'passive' building is one whose indoor environment is controlled not by the operation of a mechanical heating and cooling system but by the structure and design of the building and its components. An occupant's response to a passive, heavyweight building, which delivers very steady indoor conditions and needs only minimal heating and cooling, will be little different from their response to a mechanically controlled building, provided the constant temperature which it maintains is within the local cultural norms. The mean indoor temperature in such a passive building is decided, not by the building services engineer who designs and sets the heating and cooling systems, but by the architect manipulating the physics of the building form, orientation, mass, windows, shading, insulation, construction and materials, and passive technologies through passive building design.

Of course, the thermal conditions within passively cooled buildings are not completely unchanging. The mean temperature in the building will change from season to season, and unless it is a zero-energy building, it will often be heated or cooled for at least part of the year. In a well-designed, passively cooled (or heated) building, the building itself will moderate both the diurnal swings and the seasonal changes in indoor temperature. During hot spells a building with well-designed passive cooling provides indoor conditions that, while cooler than a building without such design features, are not likely to be as cool as would be usual with a mechanical system. So we might expect that in Figure 3.2 the comfort temperature would lie below that of a free-running building, but above that of a mechanically cooled building. So long as the passively cooled building provides temperatures between the two we can assume it will be acceptable. Such a comfort zone is incorporated in Dutch guidelines (Boerstra et al., 2002).

### 3.7.2 Mixed mode or hybrid buildings

A mixed-mode or hybrid building is one that operates in the free-running mode whenever possible, but has heating available if needed in cold weather and cooling available if needed in hot weather. In one sense nearly all naturally ventilated buildings are essentially mixed-mode, requiring some energy to heat and/or cool over the year. Certainly in northern Europe buildings are heated in the winter, even if they are not mechanically cooled in summer. So it is possible to treat the indoor climate in winter as controlled by the building thermostat and in summer by the thermal properties of the building, as well as by the actions of the occupants in both seasons.

As building technology develops, and the climate warms, this relatively simple two-mode style of operation is becoming less common because of poor standards of the climatic design and because of rising external temperatures caused by global warming and the growing urban heat island effect (see glossary). Passive and low-energy cooling systems such as shading, night-time ventilation and fans can compensate for rising temperatures up to a point, for some part of the summer in most places in Europe (Haves et al., 1998).

### 3.7.3 Traditional buildings

In many traditional buildings the thermal environment changes not only in time but also in space. Thus in an English winter the method used to keep warm might be to move closer to an intense radiant source such as an open fire. The temperature in an Indian Haveli (Figure 3.8; Matthews, 2000) or an Iranian house will vary from the ground-coupled basement to the lightweight upper floors. As they respond differently to the daily and seasonal changes in the weather, the inhabitants can move around the building to optimise their comfort.

Because the approach to comfort in modern buildings has centred on the use of air conditioning, it is assumed that unchanging conditions are more likely to be comfortable than variable conditions. There is a danger that building regulations will require traditional buildings to conform to this modern paradigm, ignoring the intentions of the original designer or builder. The adaptive approach has eroded the notion that

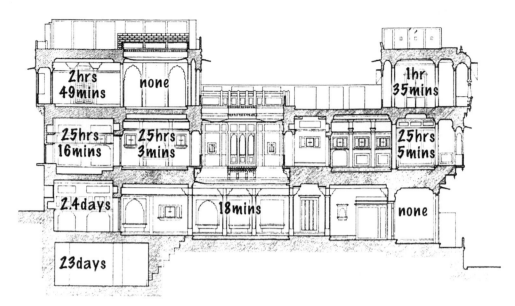

**FIGURE 3.8** Indoor temperatures display very different time lags behind outdoor temperatures depending on the extent to which they are coupled or de-coupled to the sky (fast time lag) or the earth (slow time lag) in a Haveli in Jaisalmer, India – south is to the right of this section.

Source: Jane Matthews.

unchanging conditions are superior by showing that indoor comfort can change in time according to the outdoor conditions. On the basis of the adaptive principle there is no reason why this cannot be extended to environments that change in space as well as in time. Newsham (1992) in the UK and Heidari (2010) in Iran demonstrated that building occupants do move within a space to improve their comfort. In reality a building is composed of many different microclimates and what might be true for a room facing in one direction on the ground floor may not apply to a similar room on the top floor. In understanding how buildings actually perform in terms of providing comfort, the site context is also very important. More research is needed on these questions.

## 3.8 The challenge of climate change

The changing climate poses designers with the challenge of keeping temperatures within safe ranges. We need to be able to respond adequately to climate change but with soaring energy prices this needs to be done without excessive energy use. An important factor in achieving the goal of comfort is that local populations understand the type of thermal conditions that are a risk to their health or survival. Whether a population can adapt to conditions beyond their normal experience and to what extent existing buildings will be adequate when conditions change is a matter of concern. Given the opportunity, many people are capable of adapting to changes in their own local environments. Problems of comfort may become problems of health if conditions become more variable and people are without the means, or the understanding, to deal with these adverse effects. This was tragically demonstrated in 2003 when 35,000 deaths occurred in Europe owing to unexpectedly hot weather.

At the same time the dependence of centrally controlled systems on the availability of energy can lead to complete building and system failure in extreme temperatures. At the refurbished HM Treasury building in London the entire staff were sent home by noon on 8 August during the 2003 heatwave, because the building was simply too hot to occupy. It is unclear whether this was the fault of the refurbishment which completely covered over an inner court, or whether the air conditioning was not properly commissioned. The centrally controlled system provided no opportunity for local system overrides or even to simply open a window, and this was a building that had previously been naturally ventilated successfully for nearly a century (Roaf et al., 2009, p. 59).

## 3.9 Lessons of the adaptive model for ensuring thermal comfort[2]

Thermal adaptation is essentially dynamic. The control inhabitants can exert over their environment will partly be decided by the design of building they occupy but also by the management of the building, the environment surrounding it (the need to exclude noise, dust, etc.) and so on. The indoor temperatures needed for comfort will also change with time and there are undoubtedly limits to the range of indoor climates that any group of people can adapt to over a certain period. The limits are related as much to their thermal experience as to their physiology, being affected by climate and the social, economic and cultural context.

A dynamic model for comfort requires a different approach to providing comfort than one that assumes a single temperature is best. Change and movement, typically within the context of well-understood patterns of behaviour, are the essence of the adaptive approach. Stasis, the existence of a static relationship between occupant and environment, is only achieved in circumstances where the indoor climate is highly regulated (Nicol and Roaf, 2005). To be successful, buildings need to allow people opportunities for adjustment, preferably using familiar technologies. Below are some simple guidelines for successful design and operation.

### 3.9.1 Occupant control

Adaptation is enabled by providing thermal control. Convenient and effective means of control should be provided so that the occupants can adjust the thermal environment to their own local requirements. This adaptive opportunity may be provided, for instance, by ceiling fans and openable windows in summertime, or by local temperature controls in winter. A control band of ±2K (Nicol and Humphreys, 2007) (or an equivalent range of air speed) should be sufficient to accommodate the great majority of people. Individual control is more effective than group control.

### 3.9.2 Customary thermal environments and comfort

People adapt more readily to thermal environments they are familiar with. The building should therefore be designed to provide a thermal environment that is within the range that

---

2  This section is based on the adaptive comfort section of Chapter 1 of the CIBSE Guide A (CIBSE, 2006).

is customary for the particular type of accommodation, according to climate, season and cultural context. This 'customary' temperature is usually expressed in terms of operative temperature (CIBSE, 2006). Operative temperature combines the effects of radiant and air temperature, and typical customary values can often be found in guides and textbooks, based on the experience of building services professionals.

### 3.9.3 Drift of comfort conditions

These customary temperatures are not fixed, but are subject to gradual drift in response to a changing environment both outdoors and indoors and are modified by climate and social custom. A sudden departure from the current customary temperature is likely to provoke discomfort and complaint, while a similar change, occurring gradually over several days is less likely to do so.

### 3.9.4 Dress codes

The extent of seasonal variation in indoor temperature that is consistent with comfort depends on the occupants' ability to wear cool clothing in warm conditions and warm clothing when it is cool. Strict dress codes can therefore affect thermal comfort in offices and any code should incorporate adequate seasonal flexibility and personal choice.

### 3.9.5 Temperature drift during a day

Field studies have found that in offices and schools people adjust their clothing relatively seldom during the working day (Humphreys, 1979), so the temperature during occupied hours in any day should not vary much from the comfort temperature. Temperature drifts within ±1K of the customary temperature would attract little notice; ±2K could cause mild discomfort among a small proportion of the occupants.

### 3.9.6 Temperature drift over several days

Clothing and other adjustments in response to day-on-day changes in weather and season occur quite gradually (Humphreys, 1978; Nicol and Raja, 1996, Morgan *et al.*, 2002) and take a week or so to complete. The day-to-day changes in mean indoor operative temperature during occupied hours should not normally exceed about 1K, nor should the cumulative change over a week exceed about 3K. These figures apply to sedentary or lightly active people (see also the section below on running mean temperature). If these simple suggestions are followed, people can be comfortable in naturally ventilated buildings in many climates during the whole, or part, of the year, reducing the need for air conditioning.

### 3.9.7 Predicting the most likely customary temperature from the outdoor temperature

In a survey of data from all over the world, Humphreys *et al.* (2010) found that the relationship between indoor and outdoor temperatures in free-running buildings was strong and linear. Because the people were, over time, adapted to their mean indoor temperature, their comfort temperature also followed a linear relationship with outdoor temperature.

The mean comfort temperature followed the relationship

$$T_{comf} = 0.53T_o + 13.8 \qquad \text{(standard error = 1.8K, R = 0.97)} \qquad (3.1)$$

where $T_{comf}$ is the comfort temperature and $T_o$ is the prevailing outdoor temperature (Humphreys *et al.*, 2010).

These are results from the 'average' free-running building. The fact that all buildings or occupants are not identical is reflected in the range of about 7K in the comfort temperatures at any one value of outdoor temperature. Some of the buildings measured may have incorporated some passive cooling but on the whole these were typical buildings for their location.

### 3.9.8 Time in the relationship of comfort temperatures to climate

The relationship between indoor comfort and outdoor temperature has usually been expressed in terms of the monthly mean of the outdoor temperature (Humphreys, 1981; ASHRAE, 2004). This is because the monthly mean of outdoor temperature is generally available from meteorological records. Important variations of outdoor temperature do, however, occur at much shorter intervals. Adaptive theory suggests that people respond on the basis of their thermal experience, with more recent experience being more important. A running mean of outdoor temperatures, weighted according to their distance in the past, is therefore more appropriate than a monthly mean.

### 3.9.9 Exponentially weighted running mean outdoor temperatures

An exponentially weighted running mean of the daily mean outdoor air temperature $T_{rm}$ is an appropriate expression of the outdoor temperature, and is calculated from the series:

$$T_{rm} = \{T_{od-1} + \alpha T_{od-2} + \alpha^2 T_{od-3} \ldots\}/\{1 + \alpha + \alpha^2 \ldots\} \qquad (3.2)$$

where $T_{od-1}$ is the daily mean outdoor temperature for the previous day, $T_{od-2}$ is the daily mean outdoor temperature for the day before and so on. $\alpha$ is a constant between 0 and 1 and governs how quickly the running mean responds to the outdoor temperature. The divisor is a geometric progression whose sum to infinity is $1/(1-\alpha)$, so the expression becomes:

$$T_{rm} = (1 - \alpha)\{T_{od1} + \alpha T_{od-2} + \alpha^2 T_{od-3} \ldots\} \qquad (3.3)$$

The use of an infinite series would be impracticable if equation 3.2 was not reducible to the form:

$$_nT_{rm} = (1 - \alpha)T_{od-1} + \alpha \,_{n-1}T_{rm} \qquad (3.4)$$

where $_nT_{rm}$ is the running mean temperature for day n and $_{n-1}T_{rm}$ for the previous day.

So if the running mean has been calculated (or estimated) for one day it can be readily calculated for the next day and so on.

### 3.9.10 Defining adaptive comfort in European offices incorporating time and context

In Europe extensive surveys were conducted of office workers and a value in the region of 0.8 was found to be most suitable for the constant (Nicol and McCartney, 2001; McCartney and Nicol, 2002). This value suggests that the characteristic time subjects take to adjust fully to a change in the outdoor temperature is about a week.

Equations for optimum comfort temperature were developed from the SCATs project (Nicol and McCartney, 2001) for free-running and heated or cooled modes of operation giving

Free-running:        $T_{comf} = 0.33T_{rm} + 18.8$                                    (3.5)

Heated or cooled     $T_{comf} = 0.09T_{rm} + 22.6$                                    (3.6)

Comfort temperatures are shown in relation to the running mean outdoor temperature on Figure 3.9 both for the free-running and for the heated or cooled mode. A thermally successful building is one whose indoor temperatures change only gradually in response to changes in the outdoor temperature (see above), and rarely stray beyond these bands. If we assume that people can be comfortable in a 'zone' of temperatures no more than 2K from the comfort temperature (Nicol and Humphreys, 2007) then the limits of the comfort zones for Europe (Figure 3.10) are given by the equations:

For free-running operation

upper margin:        $T_{comf} = 0.33T_{rm} + 20.8$                                    (3.7)

lower margin:        $T_{comf} = 0.33T_{rm} + 16.8$                                    (3.8)

For heated or cooled operation

upper margin:        $T_{comf} = 0.09T_{rm} + 24.6$                                    (3.9)

lower margin:        $T_{comf} = 0.09T_{rm} + 20.6$                                    (3.10)

These data have subsequently been used to underpin the comfort limits for buildings without mechanical cooling in the European standard EN15251 (CEN, 2007).

## 3.10 An example: naturally ventilated office in summer

For the assessment of the adequacy of the building in summer, the upper margin of the free-running zone is examined. This line may be used to indicate the probable upper limit of the comfort temperature (Nicol and Humphreys, 2007). In the UK the running mean outdoor temperature rarely exceeds 20°C. At this temperature the upper limit of the band is 27.4°C. So the temperature during occupied hours should preferably not exceed this value. Operative temperatures drifting a little above this value might attract little notice, but temperatures 2K or more above it would be likely to attract increasing complaint. On a more normal summer day when the running mean outdoor temperature might be 15°C, the value of the upper limit would be 25.8°C, and the indoor temperature should preferably not be higher than this. Again,

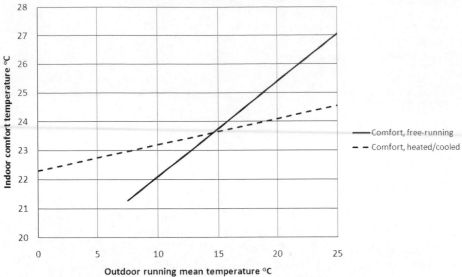
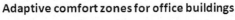

**FIGURE 3.9** Comfort temperature for buildings in free-running mode (continuous line) and heated or cooled mode (dashed line)

Source: CIBSE, 2006.

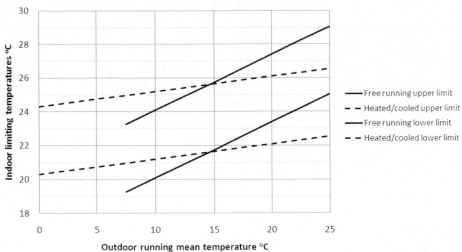

**FIGURE 3.10** Comfort zones for buildings in free-running mode (continuous lines from equation 1 ± 2K) and heated or cooled mode (dashed lines)

Source: CIBSE, 2006.

temperatures a little above this value would attract little notice, while temperatures more than about 2K above the line would be likely to attract increasing complaint.

Expected percentages of occupants experiencing discomfort have sometimes been estimated (ASHRAE, 2004; Nicol and Humphreys, 2007), but the percentage varies from building to building, depending on where its comfort temperature lies within the band, and on the adaptive opportunity it affords (Baker and Standeven, 1996). Temperatures below these values would be found satisfactory provided the advice on within-day and day-on-day temperature changes (see above) is observed.

## References

ASHRAE (2004) Standard 55-2004. *Thermal Environmental Conditions for Human Occupancy*. Atlanta, Georgia: American Society of Heating, Refrigerating and Air Conditioning Engineers.

Auliciems, A. (1981) Towards a psychophysiological model of thermal perception, *Int J of Biometeorology* 13, 147–162.

Baker, N.V. and Standeven, M.A. (1995) A behavioural approach to thermal comfort assessment in naturally ventilated buildings, *Proceedings CIBSE National Conference*, Eastbourne, 76–84.

Bedford, T. (1936) *The warmth factor in comfort at work*, Medical Research Council Industrial Health Research Board, Report 36. London: HMSO.

Boerstra, A.C., Raue, A.K., Kurvers, S.R., van der Linden, A.C., Hogeling, J.J.N.M. and de Dear, R.J. (2002) A New Dutch adaptive thermal comfort guideline. In H. Levin (ed.) *Proceedings of the 9th International Conference on Indoor Air Quality and Climate* 2, Santa Cruz: Indoor Air 2002.

CEN (2007) Standard EN15251. *Indoor Environmental Input Parameters for Design and Assessment of Energy Performance of Buildings: Addressing indoor air quality, thermal environment, lighting and acoustics*, Brussels: Comité Européen de Normalisation.

CIBSE (2006) *Environmental Criteria for Design*, Chapter 1: Environmental Design: CIBSE Guide A. London: Chartered Institution of Building Services Engineers.

Clarke, J. (2001) *Energy Simulation in Building Design*. London: Butterworth-Heinemann.

de Dear, R.J., Leow, K.G. and Ameen, A. (1991) Thermal comfort in the humid tropics – Part 1: Climate chamber experiments on temperature preferences in Singapore. Part 2: Climate chamber experiments on thermal acceptability in Singapore, *ASHRAE Transactions* 97(1), 874–879, 880–886.

de Dear, R.J. and Brager, G.S. (1998) Developing an adaptive model of thermal comfort and preference, *ASHRAE Technical data bulletin* 14(1), 27–49.

Givoni, B. and Belding, H.S. (1962) The cooling efficiency of sweat evaporation, *Proceedings of the First International Congress of Biometeorology*. London: Pergamon Press, 304–314.

Givoni, B., Khedari, J. and Hirunlabh, J. (2004) Comfort formula for Thailand, in *Proceedings of the National Solar Energy Conference (ASES) 2004*. Portland, Oregon.

Givoni, B., Khedari, J., Wong, N.H. and Feriadi, H. (2005) Responses to humidity in hot humid climates, *Proceedings of the 10th International Indoor Conference on Indoor Air Quality and Climate, 4–9 September, 2005*. Beijing.

Haldi, F. and Robinson, D. (2010) On the unification of thermal perception and adaptive actions, *Building and Environment* 45(11), 2440–2457.

Haves, P., Roaf, S. and Orr, J. (1998) The impacts of climate change on European buildings, *Proceedings of PLEA Conference, May 1998*. Lisbon.

Heidari, S. (2010) Coping with nature: Ten years thermal comfort studies in Iran, *Proceedings of Conference on Adapting to Change: New Thinking on Comfort, Cumberland Lodge, Windsor, UK, 9–11 April 2010*. London: Network for Comfort and Energy Use in Buildings. Available at http://nceub.org.uk.

Humphreys, M.A. (1973) Classroom temperature, clothing and thermal comfort: A study of secondary school children in summertime. *J. Inst. Heat. & Vent. Eng.* 41, 191–202.

Humphreys, M.A. (1976) Field studies of thermal comfort compared and applied, *J. Inst. Heat. and Vent. Eng.* 44, 5–27.

Humphreys, M.A. (1978) Outdoor temperatures and comfort indoors, *Building Research and Practice (J CIB)* 6(2), 92–105.

Humphreys, M.A. (1979) The influence of season and ambient temperature on human clothing behaviour. In P.O. Fanger and O. Valbjorn (eds) *Indoor Climate*. Copenhagen: Danish Building Research Institute.

Humphreys, M.A. (1981) The dependence of comfortable temperatures upon indoor and outdoor climates. In K. Cena and J.A. Clark (eds) *Bioengineering, thermal physiology and comfort*. Amsterdam: Elsevier.

Humphreys, M.A. (1995) Thermal comfort temperatures and the habits of Hobbits, in F. Nicol, M. Humphreys, O. Sykes and S. Roaf (eds) *Standards for Thermal Comfort*. London: Spon, 3–13.

Humphreys, M.A. and Nicol, J.F. (1998) Understanding the adaptive approach to thermal comfort. *ASHRAE Transactions* 104(1), 991–1004.

Humphreys, M.A., Rijal, H.B. and Nicol, J.F. (2010) Examining and developing the adaptive relation between climate and thermal comfort indoors. *Proceedings of Conference on Adapting to Change: New Thinking on Comfort, Cumberland Lodge, Windsor, UK, 9–11 April 2010.* London: Network for Comfort and Energy Use in Buildings. Available at http://nceub.org.uk.

Leaman, A.J. and Bordass, W.T. (1997) *Productivity in Buildings: The 'killer' variables*, London: Workplace Comfort Forum.

Levins, R. and Lewontin, R. (1985) *The Dialectical Biologist*. Cambridge, MA: Harvard University Press.

Matthews, J. (2000) The Havelis of Jaiselmer. Ph.D Thesis. East London University.

McCartney, K.J. and Nicol, J.F. (2002) Developing an adaptive control algorithm for Europe: Results of the SCATs project, *Energy and Buildings* 34(6), 623–635.

Morgan, C.A., de Dear, R.J. and Brager, G. (2002) Climate clothing and adaptation in the built environment, in H. Levin (ed.) *Proceedings of the 9th International Conference on Indoor Air Quality and Climate* 5, 98–103, Santa Cruz, Indoor Air 2002.

Newsham, G. (1992) Occupant movement and the thermal modelling of buildings. *Energy and Buildings*, 18, 57–64.

Nicol, F. and McCartney, K. (2001) *Final Report (Public) Smart Controls and Thermal Comfort (SCATs).* Report to the European Commission of the Smart Controls and Thermal Comfort project. Oxford: Oxford Brookes University.

Nicol, F. and Raja, I. (1996) *Thermal Comfort, Time and Posture: Exploratory studies in the nature of adaptive thermal comfort*. Oxford: School of Architecture, Oxford Brookes University.

Nicol, F. and Roaf, S. (2005) Post occupancy evaluation and field studies of thermal comfort. *Building Research and Information* 33(4), 338–346.

Nicol, J.F. (2004) Adaptive thermal comfort standards in the hot-humid tropics. *Energy and Buildings* 36(7), 628–637.

Nicol, J.F. and Humphreys, M.A. (1973) Thermal comfort as part of a self-regulating system. *Building Research and Practice (J. CIB)* 6(3), 191–197.

Nicol, J.F. and Humphreys, M.A. (2004) A stochastic approach to thermal comfort, occupant behaviour and energy use in buildings. *ASHRAE Transactions* 110(2), 554–568.

Nicol, J.F. and Humphreys, M.A. (2007) Maximum temperatures in European office buildings to avoid heat discomfort. *Solar Energy* 81(3), 295–304.

Preiser, W. and Vischer, V. (2004) *Post Occupancy Evaluation*. New York: Harper Collins.

Raja, I.A. and Nicol, J.F. (1997) A technique for postural recording and analysis for thermal comfort research. *Applied Ergonomics* 28(3), 221–225.

Rijal, H., Humphreys, M., Tuohy, P., Nicol, F. and Samuel, A. (2012) Considering the impact of situation-specific motivations and constraints in the design of naturally ventilated and hybrid buildings. *Architectural Science Review* 55(1), 35–48.

Roaf, S. Crichton, D. and Nicol, F. (2009) *Adapting Buildings and Cities to Climate Change*. London: Architectural Press.

Sharma, M.R. and Ali, S. (1986) Tropical summer index – a study of thermal comfort in Indian subjects. *Building and Environment* 21(1), 11–24.

Usable Buildings Trust. Available at www.usablebuildings.co.uk (September 2011).

Yun, G.Y. and Steemers, K. (2007) Time-dependent occupant behaviour models of window control in summer. *Building and Environment* 43(9), 1471–1482.

# 4

# THE HEAT BALANCE APPROACH TO DEFINING THERMAL COMFORT

One generally accepted definition of thermal comfort is that of ASHRAE:

*That state of mind which expresses satisfaction with the thermal environment.*

A number of researchers have set out to build models that mimic the physics and physiology of thermal comfort and a great number of thermal indices have resulted. Auliciems and Szokolay (2007) list some 20 major examples down the years, with others having identified as many as 80. This suggests that it is no easy task.

Some models are based on surveys of people's response to the environment using statistical analysis from field surveys and we looked at these in Chapter 3. They are sometimes called 'empirical' models. But, despite the definition of comfort given above being centred on a 'state of mind', the most common type of model is built on physics and physiology (see Chapter 2). These models try to make a considered analysis of the heat flows in the body and build a model based on physics and physiology. These are sometimes called 'rational' or 'heat balance' models and the best known among them are the Predicted Mean Vote (PMV) (Fanger, 1970) and Standard Effective Temperature (SET) (Gagge *et al.*, 1986). The PMV model is particularly important because it forms the basis for most national and international comfort standards. Here we give an outline of the thinking behind these models but we shall return to their use in standards in Chapter 5.

There is a necessary balance between the heat produced by the body and the heat lost from it, if life is to continue. But more is needed to explain the ways in which people achieve thermal comfort. One can imagine situations in which balance would occur but which might not be considered comfortable. For instance a person with a warm head and cold feet may be in thermal balance but their comfort is not assured! Again, a person who is shivering with cold might be in thermal balance, but is not comfortable. Thus it is possible to be in thermal balance but still feel too hot or cold. So the determination of comfort conditions is in two stages: first, finding the conditions for thermal balance and then, second, determining which of the conditions so defined are consistent with comfort.

## 4.1 The heat balance approach

As we have seen in Chapter 2, equations can be derived for all of the individual contributions to the heat balance equation, and they can be evaluated if the metabolic rate, the clothing resistance and the environmental parameters are known.

In the model that underlies the PMV, the condition for thermal comfort for a given person is that their mean skin temperature and sweat secretion must have a certain value at a given metabolic rate. The formulae for determining the conditions for comfort simply assume a required 'optimum' value for these two physiological variables. The data to establish optimal skin temperatures and sweat rates used in PMV were obtained entirely from climate chamber experiments. (A climate chamber is a special laboratory in which the physical variables – temperature, humidity and air speed – can be controlled.) In the classic climate chamber experiments subjects are dressed in standard clothing and given a task to do that may involve physical or mental work. Their comfort votes are then polled at intervals over periods of some hours.

During the experiments, sweat rates and skin temperatures were measured for people at various metabolic rates who considered themselves comfortable. Optimal conditions for thermal comfort were expressed by the regression lines (see Section 10.3) relating skin temperature and sweat rate to the metabolic rates from data in these experiments. In this way an expression for optimal thermal comfort was deduced from the metabolic rate, clothing insulation and the environmental conditions. The final equations for optimal thermal comfort are fairly complex. They are available in ISO 7730 (2005) and need not concern us here. The solutions to the equations have been computed and the results presented in the form of diagrams from which optimal comfort conditions (in terms of air temperature, mean radiant temperature and humidity, at different values of the air speed) can be found if the metabolic rate and clothing insulation of the design population are known.

The SET model (standard effective temperature – the other most commonly used model) also uses skin temperature as one of its limiting conditions, but uses skin wettedness (w) rather than sweat rate for the other limiting condition. Some researchers consider that this gives a more realistic feel for the effect of hot or humid conditions. The values for $T_{sk}$ and w are derived from the 'Pierce two-node' model of human physiology developed at the John B. Pierce Foundation in the United States (Gagge *et al.*, 1970). The 'two nodes' are the body's core and its peripheral tissues. This model was also calibrated in climate chamber experiments. The SET relates the real conditions to an 'effective temperature' that would give the same physiological response in people in standard clothing (0.5 clo) and metabolic rate (1.0 met) and a relative humidity of 50 per cent. This standard effective temperature can then be related to subjective response. SET was used as a thermal index in past editions of the ASHRAE Standard 55 but recent editions have used PMV rather than SET as an indicator of indoor comfort. This change remains controversial.

### 4.1.1 Predicted mean vote (PMV) and predicted percentage dissatisfied (PPD)

The PMV approach goes beyond the provision of a thermal index. The index is related not to an 'effective temperature' as is done in SET but sets out to predict the mean comfort vote on the ASHRAE scale (Table 2.1) of a group of people on the basis of six variables: the air temperature, the radiant temperature, the air speed, the humidity, the insulation of their clothing and their metabolic rate. The assumption behind the model is that the sensation expressed through the ASHRAE scale is caused by the 'physiological strain' caused by the environment (see Section 2.2). This strain is defined as 'the difference between the internal heat production and the heat loss to the actual environment for a man kept at the comfort values for skin temperature and sweat production at the actual activity level' (Fanger, 1970). It has been noted that this criterion for comfort is consistent with fixed ranges of sweat rate and skin temperature only for a fixed value of clothing insulation (Humphreys and Nicol, 1996). This extra load was calculated and related to the comfort votes of subjects in climate chamber experiments, done originally at Kansas State University in the USA and at the Danish Technical University Laboratory, using equal numbers of male and female students. These experiments and their resulting load calculation enabled PMV to predict what mean comfort vote would arise from any given set of environmental conditions for a given clothing insulation and metabolic rate. Tables of PMV are available in ISO 7730 for different environments for given clothing and metabolic rates and computer programs are available to calculate its value (e.g. Comfort Calculator).

The vote predicted is the mean value to be expected from a group of people, and PMV has been extended to predict the proportion of the group who would be dissatisfied with the environment. Dissatisfaction was defined in terms of the comfort vote. Those who voted outside the central three points on the ASHRAE scale (votes +3, +2, −2 and −3 in Table 2.1) were counted as dissatisfied. PPD expresses this as a percentage and its value is calculated from the value of PMV. Work based on field studies suggests that PPD does not reliably predict the discomfort caused by deviations from the comfort temperature in real-life circumstances of diverse activity and clothing (Humphreys and Nicol, 2002).

We have deliberately avoided a detailed description of PMV and PPD because its details are not central to the adaptive approach to thermal comfort. Those who need more detail should consult the international standard ISO 7730 (ISO, 2005) which is based on PMV or reference books such as McIntyre (1980), Parsons (2003) or, indeed, Fanger (1970).

## 4.2 Problems with the analytical approach

The heat balance approach has much to recommend it, particularly in separating out the effects of the different aspects of the thermal environment such as air movement and humidity, as well as clothing and activity. There are, however, possible sources of error in these models that need to be noted. Errors are essentially of two kinds: formulation errors and measurement errors (Humphreys and Nicol, 2000).

Formulation errors are those coming from the approximations in the way in which the environmental variables are expressed and combined to predict the comfort response. The complexity of the interaction between people and buildings leaves a lot of scope for interpretation as we have suggested in Chapters 2 and 3. The model itself has been built with a mixture of theory and measurement using a mixture of experimental approaches, both statistical and analytical. Because of these approximation errors, and because of the limitation that the subjective data were obtained from climate chamber studies, and in conditions where a steady state had been reached, it is unsurprising that the predictions based on them are far from perfect (Humphreys and Nicol, 2002).

Measurement errors are those coming from the measurement process, and are particularly evident when the environment, and people's response to it, are changeable. To predict conditions for optimal comfort requires knowledge of the clothing insulation and the metabolic rate of a group of individuals. The clothing insulation is obtained by the practitioner from tables in which clothing insulation is listed against descriptions of items or ensembles of clothing. The tabulated values of clothing insulation are determined in experiments using heated manikins. Metabolic rates are similarly obtained from tables of activities for which the appropriate metabolic rate is given. Tables of typical values of both clothing and metabolic rates are given in Tables 8.1 and 8.2. Both clothing insulation and metabolic rate are difficult to assess accurately. The measurement of the temperature and humidity can be made accurately enough, but air speeds can vary widely from time to time, and place to place, so it can be hard to measure the best value. More will be said about measurement problems in Part II.

For environmental designers who use PMV to decide what temperature to provide in a space, the model poses a number of problems. What clothing will the building occupants wear? What activity will they be engaged in (particularly where a number of activities are taking place in the same space)? Can we be confident that conditions in the building are close to those of the steady state in the climate chamber?

Field studies show that people worldwide accept a much more diverse set of thermal environments than the laboratory-based indices lead us to expect, because people have adapted to their own particular climate. If this is true, then the thermal environment standards derived from the laboratory-based models will typically only be applicable in highly serviced buildings that are capable of producing closely controlled indoor climates in a wide range of outdoor temperatures.

The variability of the indoor temperatures in buildings with no mechanical heating or ventilation renders the method very difficult to use. Without mechanical control the temperature will change continually with time. The inhabitants will open or close windows, open or close the blind, or change their clothing or their activity to make themselves comfortable. The PMV is of reduced value in these circumstances because it uses a steady-state model in a variable situation.

A heat exchange approach, as used in the PMV method, should ideally predict the outcome of the field survey, yet comfort temperatures computed from field studies are often difficult to reconcile with those calculated using the indices based on heat balance assumptions and experiments in climate chambers. From the point of view of the adaptive approach, the climate chamber is a special case of a room where the participants are unfamiliar with the space they occupy and usually lack any ability to control their conditions. If there is a behavioural element to the way people deal with discomfort then this lack of control will affect their response to the thermal environment. This could help explain the differences between the comfort temperature of a particular group of people and the temperature predicted using PMV.

## 4.3 Differences between the results from empirical and analytical investigations

A number of field surveys have been completed in which clothing and activity were estimated for the subjects at the time of the survey. Combining these with measurements of the environmental conditions it is possible to calculate comfort temperature according to the method of PMV. In ASHRAE project RP884 (de Dear, 1998), de Dear and Brager (2002) combined the data from a number of these surveys to form a database of field studies of thermal comfort from many different parts of the world.

### 4.3.1 Some limitations of PMV for variable climates

Analysing the database they found that the predicted comfort temperature using the PMV method was often significantly different from that obtained by statistical analysis of the actual comfort votes. In the variable conditions of naturally ventilated buildings, PMV can be seriously in error at values outside a band of temperature in the mid twenties Celsius (de Dear and Brager, 2002). The relationship between outdoor temperature and the comfort temperature experienced in naturally ventilated buildings is quite different from that between outdoor temperature and the comfort temperature predicted using PMV (Figure 4.1). The error results in PMV overestimating the discomfort of the occupants. In buildings with mechanical heating or cooling they found a much closer agreement between actual and PMV-based comfort temperatures. In addition to this slope effect, or following from it, there is a range effect. The range of average temperatures that people in different climates can find comfortable is greater than could be explained in terms of differences in clothing and activity alone, as previously found by Humphreys (1976) from his analysis of field-study data from around the world.

De Dear and Brager suggested that the errors in PMV may result from the expectations of the occupants being different, according to whether the building is naturally conditioned or

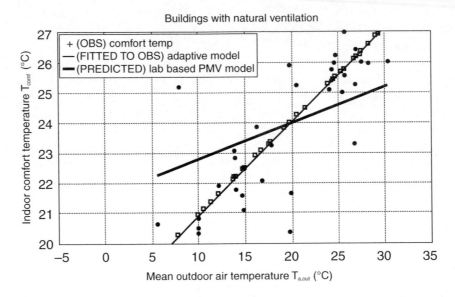

**FIGURE 4.1** The comfort temperature predicted by PMV is compared with the comfort temperature measured in actual field studies in naturally ventilated buildings. The PMV value will predict greater discomfort at high and low temperatures than is found in practice

Source: de Dear and Brager, 2002.

mechanically conditioned. There is some evidence that expectations and aspirations for comfort differ significantly between people who live a naturally ventilated lifestyle and people who live an air-conditioned one (Cândido et al., 2010) but research to explain how this works for an individual is lacking. For a comprehensive study of the discrepancies between PMV and the field study findings, see Humphreys and Nicol (2002).

### 4.3.2 Other possible sources of error

Some other possible sources of error in the predictions of the analytical method suggest themselves:

- The tables of clothing insulation used to calculate PMV assume that the insulation for a particular ensemble is independent of climate or culture. Systematic errors could occur because of a) the interpretation of the description of a particular clothing ensemble; b) climatically determined effects such as the wetting of the clothing by sweat from the skin; and c) the different ways in which clothing is actually used in different climates and cultures – loose clothing for instance is favoured in hot countries as it allows air to circulate, but its static insulation may be quite high.
- Metabolic rates are almost impossible to measure in the conditions of the field survey and so are estimated from an activity descriptor. This method may overlook some effects of climate. Metabolic rates could be affected by climate in a systematic way. To avoid discomfort in hot conditions people may carry out a particular activity in a more economical way but such an effect is not recognised by the descriptors.
- The perception of comfort in different thermal environments varies according to the order in which they happen. A mean vote predicted by a steady-state model will not reflect this. People react to an environment in a way that is to an extent dependent on their experience of it, particularly in buildings that they normally inhabit such as their home or workplace. People presented with the same thermal environment in the abnormal setting of a climate chamber will be unfamiliar with their surroundings. As a result the people in the real situation may be influenced differently by their experience, and respond according to their experience rather than according to some 'absolute' scale.

All these effects tend to increase the validity of the comfort vote in field studies compared with that of the vote predicted by the PMV model. The heat balance model is unable to take account of the social and climatic factors that are present in field surveys. Which of the effects listed is most important, and whether there are other (possibly countervailing) effects must be the subject of research and evaluation if comfort standards are to improve. In light of the above concerns, Fanger and Toftum (2002) extended the PMV model to take account of possible effects of room temperature on the metabolic rate, and of the building occupant's expectations, but the basis of this extension is speculative, and has not been adopted by ISO 7730.

### 4.3.3 Can dynamic simulation overcome the problems of steady state models?

There are now a number of dynamic physical/physiological simulation models of the human body in the context of its environment. An example is the model developed by Fiala *et al.* (1999) which has been used to develop a model of human physiological response to the outdoor climate (www.utci.org). Such simulation tools begin to address the problem of providing an adequate psychophysical model of thermal comfort responses. There are also increasingly sophisticated dynamic thermal simulation packages for buildings.

A weakness of both types of simulation, from the perspective of adaptive comfort, is that they simplify or eliminate the parts of the model dealing with the building/occupant interaction. Physiological models oversimplify the building while the building simulations oversimplify the human response, if they use a steady-state model such as PMV to evaluate the impact of the building on human comfort. Human behaviour is often inadequately treated. In order to realistically include the actual dynamic relationship between people and buildings, consideration must be given to the integration of these two very different types of dynamic model, to build a single dynamic model of the complete occupant–building system. This integration would need to be built from a solid base of appropriate empirical data.

Chapter 3 introduced an approach to human behaviour that is based on a model developed from the adaptive principle. Combining this model with a range of constraints, or other motivations afforded by the adaptive opportunities, could provide a behavioural link between buildings and occupants. This would mean a consistent model can be developed using observations of behaviour to allow designers to model the dynamic interaction between buildings and their inhabitants realistically. Such a model will also allow standardisation bodies to evaluate buildings realistically. It is evident that there is opportunity for research and development in this area.

### References

Auliciems, A. and Szokolay, S.V. (2007) *Thermal Comfort* (2nd edn). London: Passive and Low Energy Architecture (PLEA). Available at www.plea-arch.org.

Cândido, C., de Dear, R.J., Lamberts, R. and Bittencourt, C. (2010) Cooling exposure in hot humid climates: Are occupants 'addicted'? In S. Roaf (ed.) *Transforming Markets in the Built Environment: Adapting for climate change*. London: Earthscan, Chapter 6.

Comfort Calculator. Available at www.healthyheating.com/solutions.htm (September 2011).

de Dear, R.J. (1998) A global database of thermal comfort experiments. *ASHRAE Technical data bulletin* 14(1), 15–26.

de Dear, R.J. and Brager, G.S. (2002) Thermal comfort in naturally ventilated buildings: Revisions to ASHRAE Standard 55. *Energy and Buildings* 34(6), 549–561.

Fanger, P.O. (1970) *Thermal Comfort*. Copenhagen: Danish Technical Press.

Fanger, P.O. and Toftum, J. (2002) Thermal comfort in the future: Excellence and expectation. *Energy and Buildings* 34, 533–536.

Fiala D., Lomas, K.J. and Stohrer, M. (1999) Computer model of human thermoregulation for a wide range of environmental conditions: The passive system. *J. Applied Physiology (Am. Physiol. Society)* 87(5), 1957–1972.

Gagge, A.P., Fobelets, A.P. and Berglund, L.G. (1986) A standard predictive index of human response to the thermal environment. *ASHRAE Transactions* 92(2), 709–731.

Gagge, A.P., Stolwijk, J. and Nishi, Y. (1970) An effective temperature scale based on a simple model of human physiological regulatory response. *ASHRAE Transactions* 70(1), 247–260.

Humphreys, M.A. (1976) Field studies of thermal comfort compared and applied. *J. Inst. Heat. and Vent. Eng.* 44, 5–27.

Humphreys, M.A. and Nicol, J.F. (1996) Conflicting criteria for thermal sensation within the Fanger Predicted Mean Vote equation. *CIBSE/ASHRAE Joint National Conference Papers*, Harrogate, 28 September–1 October 6.

Humphreys, M.A. and Nicol, J.F. (2000) The effects of measurement and formulation error on thermal comfort indices in the ASHRAE database of field studies. *ASHRAE Transactions* 106(2), 493–502.

Humphreys, M.A. and Nicol, J.F. (2002) The validity of ISO-PMV for predicting comfort votes in everyday thermal environments. *Energy and Buildings* 34(6), 667–684.

ISO 7730 (2005) *Ergonomics of the thermal environment: Analytical determination and interpretation of thermal comfort using calculation of the PMV and PPD indices and local thermal comfort criteria*. Geneva: International Standards Organisation.

McIntyre, D.A. (1980) *Indoor Climate*. Barking: Applied Science Publishers.

Parsons, K. (2003) *Human Thermal Environments*. Oxford: Blackwell.

# 5

# STANDARDS, GUIDELINES AND LEGISLATION FOR THE INDOOR ENVIRONMENT

International comfort standards typically contain what might be termed 'Business as Usual' thinking on the indoor climate and its control. Because it is expected that energy prices will continue to rise and that the climate will continue to warm, many thermal comfort researchers are now committed to a 'Paradigm Shift' in the standards – to standards that will help drive the move to developing more resilient buildings that are capable of coping with these two potentially overwhelming challenges (de Dear and Brager, 2002; Nicol and Humphreys, 2002, 2009; Borgeson and Brager, 2011; Cândido et al., 2011). Simply tweaking the existing national and international regulations, standards and guidelines will not be enough to ensure that the necessary improvements to our building stock are made in time.

At the same time there is concern among industries that developments in those standards should help the industries' development and survival in difficult market conditions. Standards need to be flexible enough to enable building designers, engineers and managers to use every opportunity available to reduce the financial and environmental costs of achieving pleasant and healthy indoor climates. This means that standards should recognise the potential for a building, together with its occupants, to work with its installed mechanical systems to reduce energy use, and improve the health and comfort of its occupants.

## 5.1 The origin and purpose of standards for indoor climate

During the nineteenth century, research on the subject concentrated chiefly on health and safety in extreme conditions in industrial buildings and coal mines (Roaf et al., 2010). This research led to legislation to protect the health of workers in mills and factories and to developing detailed requirements for special locations such as theatres and trains. Research also addressed the high death rates among armed forces fighting in the very diverse climates of the British Empire and the Americas, and conditions in the hot, deep mines of South Africa, to improve workers' health and survival. Other standards and guidelines concentrated on vulnerable populations such as school children and hospital patients.

In the early twentieth century a new emphasis developed, as the 'miracle' of air conditioning began to open lucrative new markets to provide comfort for the emerging middle classes, in cinemas, hotels, shops and banks. Air conditioning was advertised as giving the 'zest' to dance, eat and spend, and because a refrigerating machine can produce air at almost any temperature, the research question became one of defining the temperature at which customers did those things best. By the 1930s engineers needed temperature standards to set their output air temperatures. Indoor comfort research aimed to identify these set temperatures – temperatures not about levels people needed for survival, but the levels at which they would have the stamina to work well, or spend more time and money in the shops.

At that time the environmental impacts and costs of energy use were not the driving forces. Thomas Bedford's report *The Warmth Factor in Comfort at Work* (1936) was produced by the UK Medical Research Council Industrial Health Board to address concerns about the effect of conditions in workplaces on comfort and productivity in the years during the great depression.

In his classic book on thermal comfort, Fanger (1970) identified the need for a thermal index to meet the need of the HVAC industry because

> Creating thermal comfort for man is a primary purpose of the heating and air conditioning industry, and this has had a radical influence … on the whole building industry … thermal comfort is the 'product' which is produced and sold to the customer …
>
> (Fanger, 1970, introduction)

If comfort is a product of the heating and air conditioning industry then it needs to be defined in such a way that it is possible for the industry to provide it, and to test whether the product complies with relevant standards. The approach adopted by Fanger was to devise an index (PMV) that can be precisely calculated and then use this as a proxy for comfort. This method has the advantage that the complex combination of the influences of the six key variables that influence thermal comfort (see Section 4.1) is included in a single number, the PMV. Its basis in climate chamber experimental work gives it scientific credibility.

The subsequent growing body of evidence that the PMV index did not correctly evaluate human response in buildings without mechanical conditioning (Section 4.3) had to be addressed. De Dear and Brager (2002) did this by establishing the 'adaptive' relationship between indoor comfort temperature and an index of the outdoor temperature. Again, if this index were agreed, there could perhaps again be a definition that would apply over a wide range of climates, although it is debatable whether this would result in a precise relationship with comfort. Comfort is not a precise concept and the conditions that will deliver comfort also change in a complex way (Chapter 2).

## 5.2 International comfort standards today

There are three well-known and widely used international standards that relate specifically to thermal comfort: ISO Standard 7730 (2005), ASHRAE Standard 55 (2004) and CEN Standard EN15251 (2007). All these standards are under continuous review so it is necessary to check for updates. The versions considered here are those current at the time of writing (2011).

## 5.2.1 ISO 7730

The International Standards Organisation (ISO) is the overarching source of standards, each of which becomes a national standard for its member states. ISO 7730 sets out the calculation and use of the PMV/PPD index, but also includes some criteria for local comfort. The standard has a table of measured values of the thermal insulation of various clothing items and ensembles, but does not specify what clothing people should wear. It has a table of typical values for the metabolic rates of a variety of activities. The standard specifies 'classes' or 'categories' of buildings according to the range of PMV that occurs within them: so Class A buildings maintain their indoor environment within ± 0.2 PMV (PPD ≤ 6 per cent), Class B ± 0.5 PMV (PPD ≤ 10 per cent) and Class C ± 0.7 PMV (PPD ≤ 15 per cent) (Table 5.1). The other international standards use ISO 7730 as the model, so EN15251 contains a similar categorisation. The categories are not at present included in ASHRAE Standard 55, but it is possible they will be included in future editions (Arens *et al.*, 2010). Because close control can increase energy use, the category system tends to favour high-energy buildings.

## 5.2.2 ASHRAE 55

The American Society of Heating Refrigeration and Air Conditioning Engineers (ASHRAE) controls and sponsors ASHRAE Standard 55. Because the society has numerous branches outside the USA, and because the US air conditioning industry is dominant in the international market for mechanical cooling, the ASHRAE standard is in effect an international standard. Although the standard is co-sponsored by the American National Standards Institution (ANSI) it reflects the thinking and interests of the HVAC industry, which is represented on the drafting committee. The standard is similar to ISO 7730 in being based on PMV.

The ASHRAE standard was nonetheless the first international standard to include an adaptive component. Following the extensive work of de Dear and Brager (2002) and using data collected in ASHRAE project RP884 (de Dear, 1998) an adaptive standard was developed that applies to 'naturally conditioned' buildings. The standard uses the relationship between the indoor comfort temperature and the outdoor temperature (see Section 3.4) to delineate acceptable zones for indoor temperature in naturally conditioned

**TABLE 5.1** Class A, B and C building-category specifications in ISO 7730. (The Draft Rating has been dropped from ASHRAE 55, and PMV Class A is in question because of the difficulty of measuring such small changes in PMV.)

| Category | PPD <br> Predicted percentage discomfort | DR <br> Draft rating | Local discomfort <br> (see Section 5.2.2) | PMV <br> Predicted mean vote |
|---|---|---|---|---|
| A | < 6% | < 10% | < 3–10% | −0.2 < PMV < +0.2 |
| B | < 10% | < 20% | < 5–10% | −0.5 < PMV < +0.5 |
| C | < 15% | < 30% | < 10–15% | −0.7 < PMV < +0.7 |

buildings – buildings in which the principal means of control of indoor temperature is the use of windows (Figure 5.1). The standard defines zones within which 80 per cent or 90 per cent of building occupants might expect to find the conditions acceptable. The zones are based on the comfort equation for naturally conditioned buildings derived from the RP884 ASHRAE database:

$$T_{comf} = 0.31T_o + 17.8 \qquad (5.1)$$

$T_{comf}$ is the optimal temperature for comfort and $T_o$ is the mean outdoor temperature for the survey. $T_o$ was not closely defined because surveys in the database obtained it from different sources (meteorological tables; measurements taken on site during the survey; and measurements at a nearby meteorological station during the survey).

$$T_{accept} = 0.31T_o + 17.8 \pm T_{lim} \qquad (5.2)$$

where $T_{accept}$ gives the limits of the acceptable zones and $T_{lim}$ is the range of acceptable temperatures (for 80 per cent or for 90 per cent of the occupants being satisfied). The given limits are $T_{lim}$ (80) = 3.5K and $T_{lim}$ (90) = 2.5K.

Standard 55 initially defined $T_o$ as the *monthly mean of the outdoor temperature* but it is now defined as the *prevailing mean outdoor temperature*. In the latest revision of the standard the exact choice of the form of $T_o$ is to some extent to be left to the user and can include different forms of running mean temperature. One advantage of using the monthly mean of the outdoor temperature is that it is widely available both as an historical mean over 20 or 30 years and as a monthly record. But the sudden 'jump' from one month to the next is difficult to allow for in dynamic thermal simulation. A historic monthly mean is also very insensitive to

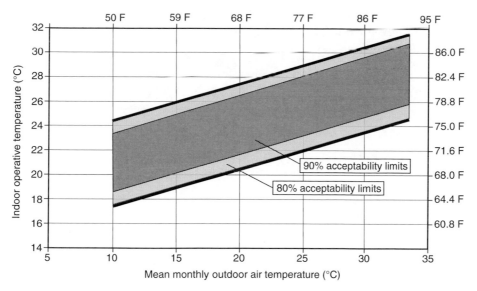

**FIGURE 5.1** Acceptable operative temperature ranges for naturally conditioned spaces

Source: ASHRAE Standard 55, 2004.

the variability of the weather from day to day and from one year to the next. There is still some question of whether the same equation can be used irrespective of what measure of outdoor temperature is used.

### 5.2.3 European standard EN15251

Standard EN15251 was developed by the Comité Européen de Normalisation (CEN) in response to calls from the European Union for standards to back up the Energy Performance of Buildings Directive (EPBD). The standard includes consideration of other aspects of the environment such as indoor air quality, lighting and acoustics as they impinge on the energy use of a building. The major thrust of the standard is the definition of the thermal environment, the sections on other factors confining themselves largely to references to other standards (Nicol and Wilson, 2011). The standard follows the general lines of the ASHRAE standard having, as well as a consideration of mechanically cooled buildings that uses PMV, an adaptive standard to be used for assessing buildings in the free-running mode.

Although EN15251 uses categories for buildings, they are defined by the nature of the building rather than referring directly to the quality of their indoor environment (Table 5.2). Mechanically conditioned and free-running buildings have the same category descriptions. Category II is recommended as the 'normal' criterion. The category descriptions in EN15251 being according to the nature of the building are an attempt to overcome the tendency of the ISO 7730 classes to favour high-energy buildings.

The adaptive standard in EN15251 is similar to that in ASHRAE 55, but using the data from the European SCATs project that was collected from five European countries instead of the ASHRAE RP884 database. Although the SCATs database contains fewer sets of comfort data, they were all collected over the same period of time in the same manner and using a standard set of instruments. The graphical form of the adaptive standard is shown in Figure 5.2. EN15251 defines acceptable values of the indoor operative temperature according to their deviation from the comfort temperature defined by the equation:

**TABLE 5.2** Category descriptions and limits for mechanically conditioned (PMV) and free-running (K) buildings in EN1525 (after Tables 1 and A1 in EN15251)

| Class/category | Description | Limitation PMV | Limitation K |
|---|---|---|---|
| I | High level of expectation and is recommended for spaces occupied by very sensitive and fragile persons with special requirements such as handicapped, sick, very young children and elderly persons | ± 0.2 | ± 2 |
| **II** | **Normal level of expectation and should be used for new buildings and renovations** | **± 0.5** | **± 3** |
| III | An acceptable, moderate level of expectation and may be used for existing buildings | ± 0.7 | ± 4 |

$$T_{comf} = 0.33\ T_{rm} + 18.8 \tag{5.3}$$

The comfort temperature is defined according to the exponentially weighted running mean of the outdoor temperature ($T_{rm}$) with the value of $\alpha$ of 0.8 (see Section 3.9.9) as this best describes the way in which comfort temperature changes with time (Nicol and Humphreys, 2010). The bandwidth of acceptable temperatures around the comfort temperature is shown in the last column in Table 5.2.

## 5.3 Discussion of international standards

In this section we discuss the strengths and weaknesses of existing standards and then later look at how changing circumstances, particularly those associated with climate change and fossil fuel depletion, will affect approaches to standards in the future.

### 5.3.1 Categories

The introduction of building categories is recent. Olesen (2010) defends their use as follows:

> The main idea behind the categories is to use them in the design of buildings and HVAC systems and to use them when evaluating the yearly performance of buildings regarding the indoor environment ... Different categories may influence the sizes and dimensioning of HVAC systems; but not necessarily the energy consumption, which is regulated through building codes and energy certification.

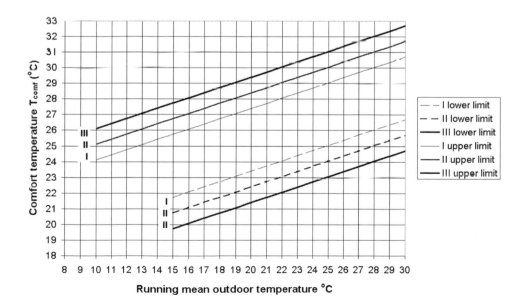

**FIGURE 5.2** Acceptable operative temperature ranges for free-running naturally conditioned spaces (after standard EN15251)

There is an underlying assumption here that all buildings will have mechanical climate conditioning systems. A principal argument against the category system is that it discourages the design and construction of buildings without mechanical conditioning. This is because without mechanical control the indoor climate is unlikely to meet the more rigorous criteria for categories A or B and hence in practice such buildings will be labelled inferior. The reality is that high-tech modern Class A buildings can be less comfortable over a year than low-tech naturally ventilated alternatives in various climates. This has been so with the many modern schools in Britain that suffer from severe overheating despite being designed to Class I/II standards, a problem not experienced in Victorian schools even in the warmer temperatures of recent decades.

### 5.3.2 Local comfort criteria

In addition to PMV, ISO 7730 incorporates a number of local thermal comfort criteria that relate to particular discomfort problems as, for example, the Draught Risk (Table 5.1). This is the supposed discomfort caused by localised air movement – maybe from an air diffuser in the air conditioning system. It addresses a real problem, for in mechanically conditioned buildings complaints of draught are common. Draughts were a problem that was assumed to apply generally. But the assumption that air movement is a problem is based on a limited perception of the role of air movement in the environment. Most people when asked say they would prefer more air movement (Zhang et al., 2010) and the Draught Risk has been found not to correlate with discomfort in field studies. In the oft-used saying 'one person's draught is another person's cooling breeze'.

Because most of the experiments underlying the 'rational approach' to comfort were conducted in climate chambers in the developed world, culturally biased assumptions can creep in unnoticed and are then difficult to dislodge. The idea that people could be comfortable at an indoor temperature of 30°C or higher, as is implied by the adaptive lines of ASHRAE 55 or EN15251, was just not considered a serious possibility, despite evidence from field studies in warm climates showing that people are indeed comfortable at these temperatures. A problem for all of us is that we do not understand our own limits until we cross them. International standards should be set by people with wide knowledge and experience of indoor temperature ranges common in climates around the world, not just those common in northern Europe or North America. Nor should their experience be confined to buildings that are air conditioned.

International Standards Agencies are addressing the cultural biases by including many local styles of dress in ISO 9920 (2009) (see Figure 2.4). Comfort is a climate- and culture-specific phenomenon and different nations are indeed developing their own unique comfort standards. Japan, China and Malaysia apply different approaches to energy savings and comfort standards developed to meet the social, cultural and economic realities of their local markets (Lau et al., 2009; Jiang and Tovey, 2009).

### 5.3.3 Mechanically conditioned buildings

The three international comfort standards use the PMV index to define acceptable internal environments. The PMV index combines the effect of four environmental and two personal variables and this makes it difficult to use, because the specification of the required value of any one variable will entail the specification of the other five. The introduction of building categories has added yet another level of difficulty.

ISO 7730 and EN15251 both give examples of the indoor temperature that will be suitable in a given set of circumstances. For instance Table A2 in EN15251 gives design minimum indoor temperatures in offices for winter (clothing insulation 1.0 clo) and maximum indoor temperature for summer (0.5 clo). Both assume a metabolic rate of 1.2 met. There are presumably assumptions about air movement (still air?) and relative humidity (50 per cent?). And 'temperature' could mean the operative temperature (or perhaps it is assumed that the air temperature is equal to the radiant temperature?). The design temperatures shown in Table 5.3 and the ranges for the categories in Table A3 incorporate all these hidden assumptions.

The advantage of such tables being included in standards is that they help the reader to judge the implications of the recommendations. The problem is that the examples can become, in effect, the accepted norm as the design temperatures. The theoretical superiority of an index that includes all the environmental and personal variables is lost because of the numerous assumptions that are made to reduce it to a room temperature.

### 5.3.4 Naturally conditioned buildings

The adaptive standards in ASHRAE 55 and EN15251 are derived from different databases. The ASHRAE standard used data from several countries and was compiled from some 30 different building surveys conducted by different research teams using different instruments. The CEN standard used data from a single coordinated group of surveys in five European countries using a uniform experimental procedure and uniform instrumentation. In addition, different methods of analysis were used (de Dear and Brager, 2002; Nicol and Humphreys, 2010). Nevertheless the two adaptive standards have almost the same comfort equation and a similar range of acceptable temperatures.

The difference in the mean outdoor temperature calculation method used in the two different standards is probably less important than it might seem. In the ASHRAE standard the outdoor temperature is expressed as the *prevailing mean outdoor temperature*, as previously explained, while in the CEN standard the equations use the exponentially weighted running mean temperature.

If one is to estimate a sensible temperature to aim for in a given climate, the historical monthly mean is a good choice. It is used in the Nicol graphs recommended in Chapter 6 to suggest climate-appropriate and comfortable indoor temperatures and to indicate the extent of the design challenge in achieving them in different climates (see Chapter 6).

TABLE 5.3 Design temperatures for offices in categories II and III from EN15251

| Category | Winter (1.0 clo) | | Summer (0.5 clo) | |
|---|---|---|---|---|
| | Design min °C | Range °C | Design max °C | Range °C |
| II | 20 | 20–24 | 26 | 23–26 |
| III | 19 | 19–25 | 27 | 22–27 |

If, however, the aim is to use building simulation to predict the indoor temperatures and compare them with the limits in the standard, then the running mean is more useful. A criticism of the exponentially weighted running mean is that people will be unfamiliar with it. However, the running mean is quite simple to explain and is easy to use. CIBSE has decided to include the exponentially weighted running mean in its standard range of weather data for simulations. It has also made an arrangement with the Meteorological Office to provide a real-time value for the running mean temperature for various places in the UK, so at least in the UK it will become more familiar.

### 5.3.5 The assumption of an isothermal indoor environment

There is an implied assumption in all standards that the PMV or comfort temperature should be uniform throughout the space being tested. For many spaces a uniform temperature is sensible but in some spaces a variable indoor climate can provide useful adaptive opportunity (Newsham, 1992), such as moving from a shady to a sunny area.

### 5.3.6 Testing compliance with a standard

Compliance testing for thermal standards can be time-consuming and costly, especially in mechanically ventilated buildings where six variables have to be measured simultaneously to a high degree of accuracy if the small deviations of PMV that are part of the standard are to be reliably estimated.

In the European project EIE-07-190: Comfort Monitoring for CEN standard EN15251, linked to EPBD (COMMONCENSE) (Nicol and Wilson, 2011), monitoring was undertaken according to ISO Standard 7726 (2001), which lays down measurement protocols to be used in defining the environment. ISO 7726 is quoted as the recommended methodology for all three comfort standards. In order to ensure that the testing occurred in representative weather for summer and winter (as suggested by the standard) the monitoring was conducted in a number of representative rooms over two four-week periods in summer and winter. The researchers at Politechnico di Milano (Zangheri, 2010) estimated that the cost of good quality testing of the instruments alone could be €10,000 (2010 prices) and would require two persons to conduct the monitoring and associated questionnaire survey, adding maybe another €3,500. Not a project that would be undertaken lightly or a service that would be cheap to provide.

Testing can also be undertaken using dynamic computer thermal simulation. This can be cheaper to perform and may increase the number of spaces within the building that can be tested. However, the difficulty in ascertaining correct values for the input data and the different predictions that result from using different simulation packages make this approach inherently less satisfactory than direct measurement.

## 5.4 Legislation

People sometimes assume that standards impose a legal duty on people to keep buildings to the temperatures specified. There is no law in any country that forbids people to open the window or adjust their indoor environment as they choose. The indoor climate for any particular building is a matter of choice by the occupant or other relevant decision maker. The concerns of the occupant must be paramount – standards have been developed to help these decision makers provide good conditions for comfort and productivity for the building occupants.

In the UK there is the Offices, Shops and Railway Premises Act of 1963 which sets a minimum temperature of 16°C for such workplaces. There was also legislation during the oil crisis in the 1970s to restrict the temperature to which buildings could be heated to 19°C, but this is largely ignored in practice. In 1961 the Parker Morris standards for the design of social housing in the UK suggested a heating system capable of maintaining 13°C in kitchens and circulation spaces and 18°C in living areas (MOHLG, 1961) but this was not compulsory until 1969. The UK Health and Safety Executive (HSE) (2011) suggest acceptable indoor environments can range from 13°C to 30°C, though they do suggest that the lower end of this range is most appropriate for places where occupants are active and the higher end for more sedentary occupations. The HSE also suggests a limitation on dissatisfaction above which action should be taken. These limits are 10 per cent for air-conditioned offices, 15 per cent for naturally ventilated offices and 20 per cent for shops.

It is hard to find legislation about maximum temperatures in buildings. Even guidelines seem to be vague. Most guidelines for offices suggest a range similar to that of the Australian Standard AS 1837-1976 (1976) which recommends a range of 21–24°C for offices. The British Council for Offices (2010) recently changed their recommendation for temperatures in offices in summer from 22°C ± 2K to 24°C ±2K following a report that showed that in summer office workers can actually be more comfortable at the higher temperature (cf. equation 3.6 in Chapter 3).

The only legal obligations that exist in the UK to maintain particular occupied indoor air temperatures come from health and safety legislation that requires, for instance, that occupants may be instructed to leave schools or workplaces under certain indoor temperature conditions. In the schools Building Bulletin 101 of the UK Department for Education and Skills (2006), the following summertime thermal criteria are specified:

- internal air temperature should be no more than 5°C above the external air temperature.
- indoor air temperature should only exceed 28°C for 120 hours or less a year.
- internal air temperature, when the space is occupied, should not exceed 32°C.

These performance standards are not enforced and are treated as advisory only.

Temperature limits that are being enforced are those in the Japanese Cool Biz programme (Enomoto *et al.*, 2009). In 2005 the Japanese Ministry of the Environment ordered all occupants of central government ministry buildings to adjust their thermostats to 28°C in the summer months. The 'Cool Biz' dress code was introduced and workers were advised to wear trousers made from materials that are air-permeable and absorb moisture, to wear short-sleeved shirts, and no jackets or ties (see Figure 5.3 in Plates). Even some of those who liked the idea of dressing more casually occasionally became self-conscious when commuting, being surrounded by non-government employees who were wearing standard business suits. Many government workers said they felt it was impolite not to wear a tie when meeting counterparts from the private sector. However, the then Prime Minister was frequently interviewed without a tie or jacket, so helping to effectively calm such fears.

The Cool Biz project has been extremely influential in Asia, and other countries are looking to follow suit. In China the building sector consumes around a quarter of all energy generated. The rate and scale of urbanisation in the country highlights the potential scale of future energy supply problems there. Jiang and Tovey (2009) calculated that increasing the thermostat temperature in summer by 1°C from 25°C to 26°C could result in energy savings in buildings of 19 per cent in Shanghai and 22 per cent in Beijing. Reducing the thermostat for the four winter months by one degree would save a further 14 per cent and 7 per cent respectively. By simply removing a jacket at work it is estimated workers can be comfortable in temperatures 2°C higher, making the appeal of the Cool Biz approach obvious. Comfort standards could then be useful in calculating how much energy can be saved by interpreting them appropriately (Sfakianaki *et al.*, 2011).

## 5.5 Standards and productivity

CEN standard EN15251 specifically mentions productivity as a reason to closely control indoor temperatures. The argument is that the cost of the extra energy used to maintain that level of control is outweighed by the cost benefits of improved productivity and the resulting decrease in staff turnover.

Productivity is notoriously hard to define and assess. Its assessment must include related indicators such as staff turnover and absence rates for sickness, as studies have shown that the push for higher productivity has often resulted in poor morale, high staff turnover, ill health (Bergqvist *et al.*, 1995) and high energy costs too.

One argument used to persuade designers and clients of the need for tight control of indoor air temperatures rests on the assumed rise in productivity associated with it. Recent work on the relationship between temperature and task performance published by REHVA (2007) rests chiefly on a meta-analysis by Seppanen and Fisk (2006) that combines numerous studies of the performance of different kinds of work, but chiefly that of telephone call centres, where the performance criterion was the time taken to answer an enquiry. They found an optimum performance at around 22°C, but with a wide band of uncertainty. The meta-analysis awaits a thorough critique, and its relevance to office work in general is uncertain, chiefly because variation in the performance of single attributes of single tasks bears only a loose relationship to the wider concept of productivity in industry.

There is a danger that the relation between room temperature and performance found in that study will be interpreted simplistically. The relation is indirect: performance, in its thermal aspect, depends on the thermal state of the body rather than on that of the room. It follows that, as people adapt to their room temperatures by, for example, seasonal changes of clothing, the room temperature for optimum performance of a particular task will show a corresponding change. Further, different kinds of task show optima at different body thermal states. Thus, it is probable that there is no single optimum temperature for task performance. Performance is therefore likely to be best at the temperatures the workers find comfortable.

This conclusion has been confirmed in a recent study in the Netherlands by Leyten and Kurvers (2010), who investigated the assumptions on which the Rehva Guide No 6 – Indoor Climate and Productivity in Offices – was based. The guide states that its main purpose is to establish quantitative relationships of indoor environmental aspects with productivity and sickness absenteeism. The Dutch study found that the following relationships could be clearly identified:

- temperature with productivity
- ventilation with productivity
- perceived indoor air quality with productivity
- ventilation with sickness absenteeism.

The purpose of their work was to establish what in practice are likely to be the most effective measures to increase productivity and decrease absenteeism, given all available evidence. They argued that robust measures, such as avoidance of indoor air pollution sources, minimising external and internal heat loads, using thermally effective building mass, cellular office layout, occupant control of temperature, operable windows and providing for adaptive thermal comfort are more effective in increasing productivity and reducing absenteeism than less robust measures such as diluting indoor air pollution through increased ventilation or controlling temperature through mechanical cooling.

Who dictates the work conditions is also a key consideration, because the ability to control one's own working conditions has been shown to improve perceived levels of comfort and in turn how productive people feel they are (Bordass and Leaman, 2007). This is borne out by the findings of a study of typical office workers across a wide range of European locations from Sweden to Portugal and Greece. Inspection of self-assessed productivity found that, over a wide range of room temperatures, people rated themselves most productive when they were most comfortable – and the temperature for comfort varied widely between the different countries and climates of Europe (Humphreys and Nicol, 2007).

## 5.6 Standards and overheating in buildings

Overheating in buildings is a condition that most people will recognise. Avoiding overheating has been the motive for a considerable number of research projects. CIBSE have traditionally defined an overheating building as one where the temperature exceeds a certain value (e.g. 28°C) for a specified length of time (e.g. 1 per cent of working hours). Nicol *et al.* (2009) suggest that this definition of overheating is flawed for a number of reasons:

- Although overheating can occur in mechanically cooled buildings, the solution to the problem there is managerial rather than environmental. The greatest danger of overheating is in buildings in the free-running mode whose poor design has meant higher temperatures. For free-running buildings, as we have seen in Chapter 3, temperature limits are adaptive and a single temperature limit will give a poor estimate of the overheating danger.
- An 'hours over' criterion can define the frequency of overheating but not the severity: 8 hours over the limit by 1K may be more acceptable than 4 hours by 3K. Overheating is a function of time period and severity.
- The 'hours over' criterion and any criterion that has a sharp threshold temperature will be very sensitive to the nature of the assessment method for internal temperatures. Systematic errors in calculation methods can lead to changes in the shape of the distribution of indoor temperatures that can significantly affect the extreme 'tail', which is the critical part of the distribution for discomfort.
- Even if the likelihood of discomfort is a good measure of overheating, the perception that a building overheats may be more subtle and include other factors such as room

size, the effectiveness of the ventilation, the number of occupants and so on. Careful and perceptive research is needed from which a definition of overheating can be developed.

Future standards will need to provide a definition of overheating which, while not complete, will at least overcome the shortcomings of the 'hours over' model. Nicol and Humphreys (2007) have shown that the likelihood of discomfort in European free-running buildings is a function of difference of the actual indoor temperature from the comfort temperature as defined by equation 5.3. The ASHRAE 55 adaptive standard also predicts the likelihood of discomfort as shown in equation 5.2 for 90 per cent or 80 per cent acceptability limits.

## 5.7 The way forward for comfort standards

This short introduction to indoor temperature standards and the thinking behind them suggests that the existing standards are not well suited to encouraging robust, durable approaches to the design of indoor climates in an era of soaring energy prices and rising global temperatures. Such concerns have long been voiced (Nicol and Humphreys, 2002). The complexities of the indoor environment mean that standards based on temperature, humidity, clothing and so on are not easy for the building designer to work with. Standards that talked in terms of windows, walls and doors, and their physical placement and construction would be more likely to be understood by designers and the construction industry.

A comfort standard that seeks to limit energy use rather than being centred on an index of the environment has been suggested by Nicol and Humphreys (2009). They suggest that classifying buildings according to their indoor climate is inappropriate. A more useful way would be to define how good a building was by how little energy it used in order to provide adequately comfortable conditions for the uses to which that building was put. Thus, a Class A building might be one that could remain comfortable with no use of mechanical energy and a building that was profligate in its energy use would be considered as Class C. While comfort remains an essential requirement of the standard, it can be defined in an adaptive way such as is suggested in Section 3.8 above. There remain problems with such new approaches to the framing of standards and clearly change cannot be achieved quickly, but the inappropriateness of current approaches is becoming ever clearer, and new thinking is overdue.

## References

Arens, E., Humphreys, M.A., de Dear, R.J. and Zhang, H. (2010) Are 'class A' temperature requirements realistic or desirable? *Building and Environment* 45(1), 4–10.

AS 1837-1976 (1976) Code of practice for application of ergonomics for factory and office work. Sydney: Standards Australia.

ASHRAE 55 (2004) Standard 55-2004. *Thermal Environmental Conditions for Human Occupancy*. Atlanta, Georgia: American Society of Heating, Refrigerating and Air Conditioning Engineers.

Bedford, T. (1936) *The Warmth Factor in Comfort at Work*. Medical Research Council Industrial Health Research Board, Report 36. London: HMSO.

Bergqvist, U., Wolgast, E., Nilsson, B. and Voss, M. (1995) Musculoskeletal disorders among visual display terminal workers: Individual, ergonomic, and work organizational factors. *Ergonomics* 38(4), 763–776.

Bordass, B. and Leaman, A. (2007) Are users more tolerant of 'green' buildings? *Building Research and Information* 35(6), 662–673.

Borgeson, S. and Brager, G. (2011) Comfort standards and variations in exceedance for mixed-mode buildings. *Building Research and Information* 39(2), 118–133.

British Council for Offices (2010) *The 24 Degree Study: Comfort, productivity and energy consumption – a case study*. London: British Council for Offices.

Building Bulletin 101 (2006) *Ventilation of school buildings*. London: Department for Education and Skills.

Cândido, C., Lamberts, R., de Dear, R.J., Bittencourt, L. and de Vecchi, R. (2011) Towards a Brazilian standard for naturally ventilated buildings: Guidelines for thermal and air movement acceptability. *Building Research and Information* 39(2), 145–153.

CEN (2007) Standard EN15251. *Indoor Environmental Input Parameters for Design and Assessment of Energy Performance of Buildings: Addressing indoor air quality, thermal environment, lighting and acoustics*. Brussels: Comité Européen de Normalisation.

de Dear, R.J. (1998) A global database of thermal comfort experiments. *ASHRAE Technical data bulletin* 14(1), 15–26.

de Dear, R.J. and Brager, G.S. (2002) Thermal comfort in naturally ventilated buildings: Revisions to ASHRAE Standard 55. *Energy and Buildings* 34(6), 549–61.

Enomoto, H., Ikeda, K., Azuma, K. and Tochihara, Y. (2009) Observation of the thermal conditions of the workers in the 'Cool Biz' implemented office, National Institute of Occupational Safety and Health. *Japan (JNIOSH)* 2(1), 5–10.

Fanger, P.O. (1970) *Thermal Comfort*. Copenhagen: Danish Technical Press.

HSE (2011) Health and Safety Executive website. Available at www.hse.gov.uk/temperature (September 2011).

Humphreys, M.A. and Nicol, F. (2007) Self-assessed productivity and the office environment: Monthly surveys in five European countries. *ASHRAE Transactions* 113(1), 606–616.

ISO 7726 (2001) *Ergonomics of the Thermal Environment: Instruments for measuring physical quantities*. Geneva: International Standards Organisation.

ISO 7730 (2005) *Ergonomics of the Thermal Environment: Analytical determination and interpretation of thermal comfort using calculation of the PMV and PPD indices and local thermal comfort criteria*. Geneva: International Standards Organisation.

ISO 9920 (2009) *Ergonomics of the Thermal Environment: Estimation of thermal insulation and water vapour resistance of a clothing ensemble*. Geneva: International Standards Organisation.

Jiang, P. and Tovey, N.K. (2009) Opportunities for low carbon sustainability in large commercial buildings in China. *Energy Policy* 37, 4949–4958.

Lau, L.C., Tan, K.T., Lee, K.T. and Mohamed, A.R. (2009) A comparative study on the energy policies in Japan and Malaysia in fulfilling their nations' obligations towards the Kyoto Protocol. *Energy Policy* 37, 4771–4778.

Leyten, J.L. and Kurvers, S.R. (2010) Robust design as a strategy for higher workers' productivity: A reaction to Rehva Guide No. 6. *Indoor Climate and Productivity in Offices, Proceedings of Conference on Adapting to Change: New Thinking on Comfort, Cumberland Lodge, Windsor, UK, 9–11 April 2010*. London: Network for Comfort and Energy Use in Buildings. Available at http://nceub.org.uk.

MOHLG (1961) *Homes for today and tomorrow*. Ministry of Housing and Local Government, London: HMSO.

Newsham, G. (1992) Occupant movement and the thermal modelling of buildings. *Energy and Buildings* 13, 57–64.

Nicol, F., Hacker, J., Spires, B. and Davies, H. (2009) Suggestion for new approach to overheating diagnostics. *Building Research and Information* 37(4), 348–357.

Nicol, F. and Wilson, M. (2011) A critique of European Standard EN15251: Strengths, weaknesses and lessons for future Standards. *Building Research and Information* 39(2), 183–193.

Nicol, J.F. and Humphreys, M.A. (2002) Adaptive thermal comfort and sustainable thermal standards for buildings. *Energy and Buildings* 34(6), 563–572.

Nicol, J.F. and Humphreys, M.A. (2007) Maximum temperatures in European office buildings to avoid heat discomfort. *Solar Energy* 81(3), 295–304.

Nicol, J.F. and Humphreys, M.A. (2009) New standards for comfort and energy use in buildings. *Building Research and Information* 37(1), 68–73.

Nicol, J.F. and Humphreys, M.A. (2010) Derivation of the equations for comfort in free-running buildings in CEN Standard EN15251. *Buildings and Environment* 45(1), 11–17.

*Offices, Shops and Railway Premises Act* (1963) London: HMSO.

Olesen, B. (2010) Why specify indoor environmental criteria as categories? *Proceedings of Conference on Adapting to Change: New Thinking on Comfort, Cumberland Lodge, Windsor, UK, 9–11 April 2010.* London: Network for Comfort and Energy Use in Buildings. Available at http://nceub.org.uk.

REHVA (2007) *How to integrate productivity in life-cycle cost analysis of building services*, P. Wargorcki and O. Seppänen (eds) Report 6 of the Federation of European Heating, Ventilating and Air Conditioning Associations (REHVA), Brussels.

Roaf, S., Nicol, F., Humphreys, M.A., Tuohy, P. and Boerstra, A. (2010) Twentieth century standards for thermal comfort: Promoting high energy buildings. *Architectural Science Review* 53(1), 65–77.

Seppanen, O. and Fisk, W.J. (2006) Some quantitative relations between indoor environmental quality and work performance or health. *International Journal of HVAC&R Research* 12(4), 957–973.

Sfakianaki, A., Santamouris, M., Hutchins, M., Nicol, F., Wilson, M., Pagliano, L., Pohl, W., Alexandre, J.L. and Freire, A. (2011) Energy consumption variation due to different thermal comfort categorization: Introduced by European Standard EN15251 for new building design and major rehabilitations. *International Journal of Ventilation* 10(2). Available at www.ijovent.org.uk.

Zangheri, P. (2010) Assessing thermal comfort in practice: Long-term metering and occupant surveys in four case studies. *Next Building: Advances in next generation building technologies and design.* Verona: SolExpo Fair.

Zhang, H., Arens, E. and Pasut, W. (2010) Thresholds for indoor thermal comfort and perceived air quality. *Proceedings of Conference on Adapting to Change: New Thinking on Comfort, Cumberland Lodge, Windsor, UK, 9–11 April 2010.* London: Network for Comfort and Energy Use in Buildings. Available at http://nceub.org.uk.

# 6

# LOW-ENERGY ADAPTIVE BUILDINGS

The third volume in our trilogy will show how to design buildings with adaptive thermal comfort in mind, and this chapter introduces some of the threads that will run through that book.

The need to understand how buildings and people interact to provide comfort is important, not just to those involved in thermal comfort research or standards, but also to those who design and occupy buildings. The comfort approach that informs a building design, the ways in which the building is used, and how it is adapted by and for its occupants, all have a significant impact on the amount of energy consumed in that building and the greenhouse gases it generates. (Greenhouse gases are any gases in the atmosphere that tend to warm the planet. Carbon dioxide is important among them, and is produced by burning coal, oil and natural gas.)

In Section 4.2 we saw that standards based on the heat balance approach are typically only applicable in highly serviced buildings. Consequently designers relying solely on such standards will be driven to design highly serviced buildings as they attempt to guarantee a fixed indoor temperature. The scale of the consequent energy use is enormous and is reflected in US energy consumption patterns. In 2010 around 40 per cent of all energy in the US and 76 per cent of electricity[1] was used in buildings (DOE, 2011). About 40 per cent of primary energy used in buildings is used for heating, cooling and ventilation. If every building were naturally ventilated during those seasons when the outdoor air was considered locally to be comfortable, large amounts of greenhouse gas emissions might be saved. Of course this cannot happen if the windows do not open and if people do not dare to open their windows even if they could (Klinenberg, 2002).

Air conditioning is essential in the hottest regions and times of year, and will increasingly be so in a warming climate. One tragedy of the twentieth century was that cheap energy meant that air conditioning too often led to poor understanding of climatic

---

1 US electricity consumption by sector: building operations, 76 per cent (residential 39 per cent, commercial operations 36 per cent); industry, 23 per cent; transport, 1 per cent.

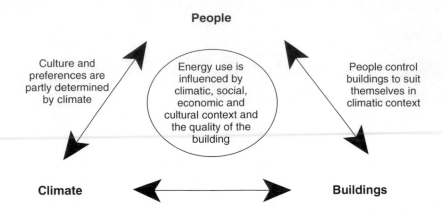

**FIGURE 6.1** To successfully design for comfort in a free-running, low-energy building requires
an understanding of the three-way interaction between climate, people and buildings

Source: Nicol from Roaf *et al.*, 2011.

building design. The reality of the energy-constrained twenty-first century (Roaf *et al.*,
2009) will be that it is necessary to have highly energy-efficient heating and cooling
systems working in high-quality passive buildings to maintain comfort during extreme
weather events. Higher energy use during extreme seasons may be compensated for by
expanding the seasons in which buildings are free running, being conditioned using the
freely available natural energy flows of the sun, earth and air around buildings, and by
running buildings on integrated renewable energy systems when possible. When this is
insufficient to maintain comfort then Thermal Energy Storage (TES) systems come into
play to reduce the peaks in energy demand that are driving power plant construction
across the world.

The impacts resulting from the poor-quality thermal design of many US buildings
are even higher. The worst problem they pose is that of the 'peak' demand that buildings place on a utility's generating capacity. The large industrial customers (over 1MW)
know how to manage their plants to control peak loads. Unfortunately, residential and
commercial customers do not have this same level of understanding. Most large US
utilities show a peak demand profile in which buildings can be 80–90 per cent of that
peak demand. Peak demand problems are driving new power plant construction, and
buildings are the major factor in creating this problem. If the US could flatten out their
peak demand they would not have to build a new power plant for the next 50 years
(Bryan, 2011), and the dependence, bordering on addiction (Cândido *et al.*, 2010), on air
conditioning in some countries makes the buildings in them, and their occupants, very
vulnerable to failure. It is the poor quality of the buildings that creates the real risk, not
the machines or the people in them that often take the blame and are seen as the culprits.
Nor is the problem confined to the US, but is true of many developed countries around
the world.

In this chapter we explore how the adaptive approach to comfort may help us design better buildings.

## 6.1 Building design and adaptive comfort

Key to mastering the skill of producing comfortable, low-carbon buildings is to look at the 'whole system' when designing. The adaptive comfort approach helps us here because it works with many attributes of the system. The outside climate, the building's context, its form, services and occupants as well as the seasons and times of day are all part of this complex package of attributes that determine our comfort in a system that is:

1   *Dynamic and interactive*: The process of adapting to different thermal environments is essentially dynamic. Understanding that the buildings we design are part of this dynamic system necessitates a different approach to achieving comfort than one that assumes that the buildings we occupy are inert boxes that are supplied with 'comfort' at a single steady temperature through a duct in the wall, floor or ceiling. Not that one approach to comfort is wrong and the other right, but rather that the different comfort approaches require different palettes of strategies, both for building design and for building services, to achieve comfort.

2   *Changing*: Change and movement, typically within the context of well-understood patterns of behaviour, are intrinsic to the adaptive approach. Change can include changes of location for different activities, as well as changes in the building itself over time. It includes movement between buildings, between rooms, around rooms, out of the sun, into the breeze, closer to the fire, with the blinds shut, with the curtains open and so on.

3   *Customary*: People who regularly occupy a particular space will have a customary temperature that they associate with that place for a particular time of day or year. It will be part of a thermal pathway they follow each day, a pathway that at times may be too hot or too cold, but on average constitutes a well-understood pattern for a generally comfortable life. These customary temperatures will change over time in response to the weather and other changing circumstances, but it is important that the temperature provided by a building is close to the customary temperature, and that within-day and between-day changes of temperature are small, as is suggested in Section 3.9.

   Changes in customary temperatures can occur in air-conditioned buildings too. Studies have shown that even in a fixed-window, climate-conditioned building, the use of an adaptive algorithm in the controls that allows indoor temperatures to track outdoor temperatures (higher in summer, lower in winter) will provide a similar degree of comfort as would a single temperature setting, while making considerable energy savings for the building manager (McCartney and Nicol, 2002).

   Although the constant or gently changing indoor temperature promised by air-conditioned buildings is one way to provide occupant comfort, it is by no means the only way. Many traditional buildings in colder climates relied on point sources of heat from fires, stoves and on hot water bottles to provide localised warmth while the background temperatures in the room were well below usual 'comfort temperatures' (Humphreys *et al.*, 2011). Sharp temperature differences were accepted and even enjoyed, when set against this well-understood behavioural pathway.

4   *Seasonally adjusted*: The customary temperature in naturally ventilated buildings changes with outdoor temperature in a more or less linear relationship when the buildings are in free-running mode – the rooms are much warmer in summer than in winter. CEN

Standard 15251 (2007) gives a range of comfortable indoor temperatures related to the running mean of the outdoor temperature. The Nicol graph (see Section 6.3.1) is another approach. The graph can act as a guide for the development of naturally ventilated build-ings that are comfortable yet do not require the high-energy use and other disadvantages of air-conditioned buildings, such as the loss of control by individuals of their own com-fort through being unable to open a window to welcome in a cooling breeze. Very passive, heavyweight buildings can include areas that maintain almost constant indoor temperatures, rather like air-conditioned buildings, and remain perfectly comfortable, being decoupled from the rising and falling outdoor temperatures (see Section 3.7.1).

5   *A goal not a product:* Comfort has been seen as a 'product' provided by the building for its occupants, as defined by Ole Fanger in 1970 (see Section 4.1). This is a premise favoured by those who sell 'comfort' as a product in the form of mechanical systems that are installed, the control systems that run them and the services associated with both of these. It is also

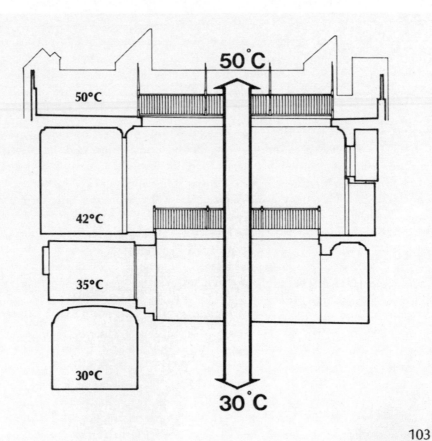

103

**FIGURE 6.2** The massive traditional houses of Baghdad couple the stable cooler temperatures of the earth to the extensive basements that are occupied during the intense heat of the summer day in a hot dry climate

Source: Susan Roaf.

a premise that is favoured by a generation of architects who are happy to assign the risk of performance failures of their buildings to the mechanical system installers and designers. This is not least because many schools of architecture no longer teach climatic design in their curricula in any depth, resulting in a deskilled generation of poor building performance designers. Unfortunately, the traditional building service engineer is also often not taught how to design low-energy buildings, or about natural ventilation, or about the use of renewable energy technologies to lower carbon emissions from buildings.

The comfort-as-product premise is, however, not so popular with a growing generation of building occupants who find themselves with no control over their own comfort conditions. The engineers who operate the 'intelligent' building management system can be 50 miles away or, even if they are in the building, cannot be contacted. All too often, even the experts find themselves unable to effectively control the often extremely complex and conflicting systems (see Usable Buildings Trust).

ASHRAE defines comfort as 'a state of mind' (Chapter 4). It is not a product for sale. Achieving comfort is a *process*. Comfort is a goal that is sought, and can be achieved by occupants, provided they are able to control their own environment. The control they can exert will partly be decided by the building they occupy, its services and the adaptive opportunities offered by each. In many workplaces management have a key role to play, and the importance they assign to comfort, the provision of avenues of complaint and adjustment, the use of dress codes and so on are essentially management decisions that can crucially affect occupant comfort. There are undoubtedly limits to the range of indoor climates that any group of people can adapt to willingly, related as much to their thermal experience and climatic, social, economic and cultural contexts as to their physiology (Humphreys and Nicol, 1998; Nicol and Humphreys, 2002).

## 6.2 Historic flaws with the mechanical approach to providing comfort

Air conditioning was originally used in hotels, cinemas, banks, factories and schools where it was believed to increase productivity, attract customers and provide better working conditions for the young and old alike. Mechanical systems of environmental control in buildings became pervasive in the United States as the technologies spread from the workplace to the home. Ackerman (2002) writes 'For better and for worse, our world tomorrow will be air-conditioned'. This is not least because the standards demand it, even in climates where the need for air conditioning is minimal (Haves *et al.*, 1998). Historically, mechanically conditioned buildings were often:

- disconnected from their surrounding climates and environments by unopenable 'envelopes';
- poorer constructions because the mechanical systems enabled occupants to be comfortable in cheaper buildings;
- energy hungry, driving climate change through high levels of greenhouse gas emissions.

Elizabeth Shove (2003) shows how air conditioning has gone from being a luxury for the few to being a necessity for the many, especially in the United States. One problem is what Shove calls the 'ratchet effect', a reference to the irreversibility of such changes – or at least the great difficulty of reversing them. The rapid move to more mechanically conditioned buildings over the last half century was in part a reaction to the decrease in standards of environmental performance of buildings themselves as architects strove to follow the tenets of 'Modern Movement' design. Large areas of glass were seen 90 years ago as important to let in healthy light and sunshine. But they resulted in glazed walls that seldom opened and often caused overheating. Buildings raised on pilottis to allow street life to flourish beneath them became wind tunnels beneath cold- or heat-exposed concrete floors (Roaf *et al.*, 2009). These failings could be readily compensated for by putting more servicing into buildings in a age when energy was talked about as being 'too cheap to meter', but such buildings are now a liability as the drive is now towards 'low-carbon buildings'.

Already in the 1960s researchers at the UK Building Research Establishment saw that the trend towards over-glazed, lightweight buildings was causing overheating, discomfort and high-energy running costs (Loudon and Keighley, 1964; Loudon and Danter, 1965; Black and Milroy, 1966; Loudon, 1968). Accompanying this trend was the growing power of the building services engineer in the design team, who was needed increasingly to avoid the risk of creating internal climates that people cannot occupy. Standards followed design trends and increasingly locked the design team into the need to use extensive and complex mechanical equipment. Low-energy, low-carbon buildings fare badly within a regulatory system that promotes mechanical solutions that can provide guaranteed indoor temperatures, regardless of their energy costs (Touhy, 2008).

A question that must be asked is why this 'ratcheting up' of poor design performance was not acted upon earlier. For some there was little impetus to interfere with a business-as-usual approach that favoured their own industries. For architects, however, a flaw in their working patterns is that they so seldom return to a building, once built, to learn from the strengths and weaknesses of the strategies used in it (Baird, 2010). Professional bodies are currently looking again at a requirement for designers to revisit buildings as part of their scope of work, to assist in correcting this flaw.

Today a key driver for change is the rising cost of fossil fuel energy. In business, energy was seldom discussed at boardroom level a decade ago when staff costs dominated the operational balance sheet. Energy costs for businesses, not least to run servers and computers, are now becoming a critical issue. Escalating fuel prices in the domestic sector are pushing whole sections of the global population into fuel poverty, as even some of the middle classes struggle to pay their heating and cooling bills. At the same time the climate is becoming more extreme, requiring more energy to heat and cool buildings to maintain safe and comfortable indoor temperatures.

However, things are changing fast in response to these challenging drivers and to the need to reduce greenhouse gas emissions to mitigate climate change. A new generation of architecturally engineered buildings, technologies and design strategies is being developed including both modern engineering solutions and also more traditional approaches such as the use of thermal mass (Figure 6.3). These buildings will work with the climate to provide comfort in buildings, while reducing their energy use and carbon penalties. These new solutions and markets require a 'whole system thinking' approach, putting the building, the occupants and the controls into the design mix, so providing locally appropriate buildings. Natural ventilation will be used when and where possible in mixed mode buildings (Ward *et al.*, 2012), using a wider range of technologies including ceiling fans (Aynsley, 2012), relying increasingly on natural energy from the sun, wind, air and earth, and new building control systems that can accommodate both mechanical systems and occupant window-opening behaviours

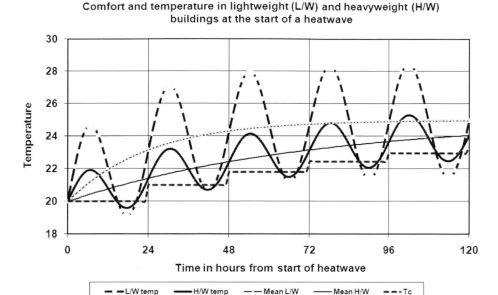

Comfort and temperature in lightweight (L/W) and heavyweight (H/W) buildings at the start of a heatwave

**FIGURE 6.3** Graph showing how the temperature varies from day to day in a heavyweight (lower curve) and a lightweight (upper curve) building in response to a step change in the weather. As well as a smaller variation in temperature each day, the mean temperature, which will eventually end up the same in both buildings, rises to its final value more slowly, allowing occupants time to adapt to the changed conditions (see comfort temperature $T_c$). So the use of well-designed thermal mass extends the adaptive potential of a building significantly, enabling it to keep occupants comfortable when external temperatures change

Source: Fergus Nicol.

(Rijal *et al.*, 2012). At the heart of these new solutions and markets is the adaptive approach to thermal comfort.

Many different tools, as well as technologies and strategies, are used to design comfortable low-carbon buildings but the quality of the result is in part dependant on the questions asked of these tools. The first question must be: how do I design a building that is optimally suited to this climate? For instance it is not sensible to put a lightweight, highly glazed building with a heat recovery ventilation system in a hot climate, whereas it might be suited to a sunny alpine one. A good method of getting an early understanding of the scale of the climatic design challenge, and a feel for what is the right building and technology type for a particular climate, is to use a Nicol graph.

## 6.3 Designing more appropriate buildings

### 6.3.1 The Nicol graph

The Nicol graph is a simple and effective tool for understanding how much heating or cooling is required to provide comfort in a building in a particular place, for local populations.

The Nicol graph (Nicol et al., 1994) starts from the adaptive tenet that the temperature that people find comfortable indoors varies with the mean outdoor temperature. This is especially true for people in buildings that are 'free running' (i.e. not mechanically heated or cooled). Figure 3.2 (from Humphreys, 1978) illustrates the relationships.

Each point in Figure 3.2 is the value of the comfort temperature determined from a survey of thermal comfort plotted against the mean outdoor temperature at the time of the survey. People are assumed to be familiar with the conditions they encounter culturally and climatically. The mean of the daily outdoor temperature $T_{om}$ is estimated from meteorological records as:

$$T_{om} = (T_{omax} + T_{omin})/2 \tag{6.1}$$

where $T_{omax}$ is the monthly mean of the daily maximum outdoor temperature and $T_{omin}$ is the monthly mean of the daily minimum outdoor temperature at the time of the survey ($T_{omax}$ and $T_{omin}$ are usually available from meteorological records and can be found on the BBC weather website for many cities worldwide). In free-running buildings the comfort temperature varies linearly with the outdoor temperature as shown in Figure 3.2. Analysis of recent data (Humphreys et al., 2010) suggests that this equation for this line should be

$$T_{comf} = 0.53 \, (T_{om}) + 13.8 \tag{6.2}$$

Using equation 6.2 a comfort temperature can be calculated for each month of the year (see the appendix to this chapter).

The relationship between comfort temperature and the outdoor temperature can be used to help design comfortable free-running buildings. Examples for a variety of climatically distinct areas of Iran where buildings are free running for a major part of the year are shown in Figure 6.4. Here the indoor comfort temperature $T_{comf}$ is calculated from the monthly value of $T_{om}$ (equation 6.1) and plotted month by month together with $T_{om}$, $T_{omax}$ and $T_{omin}$. Such a diagram helps the designer to judge whether passive heating and/or cooling is a possibility in the climate under consideration.

Givoni (1991) established that an effective passive building, well designed and with high levels of mass, could maintain a mean indoor temperature that is around the mean outdoor temperature ($T_{om}$), half way between the outdoor maximum and minimum temperatures. Thus the Nicol graph gives an easy first approximation of the cooling or heating challenge faced by the designer in that climate. The relationship between the desired indoor temperature ($T_{comf}$) and the range of outdoor temperatures shows whether, for instance, night cooling is likely to be a viable way to keep a passive building comfortable in summer, evident if $T_{om}$ and $T_{comf}$ approach each other in summer. If a large heating requirement in winter is indicated by the divergence of these two values, the extension to the graph (Roaf et al., 2011) to include the mean solar intensity will give designers a means to assess the extent to which passive solar heating may be a successful heating option in the building.

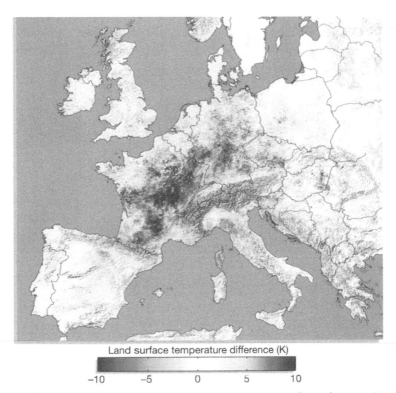

Land surface temperature difference (K)

−10    −5    0    5    10

**FIGURE 1.1** Profile of an overheating event: June–August 2003 anomalies (relative to 1961–90 mean in UK) over Europe

Source: Stott, P. *et al.* (2004) Human contribution to the European heatwave of 2003, *Nature*, 432(7017): 610–14

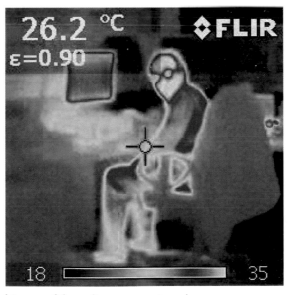

**FIGURE 2.1** A thermal image of the radiant temperatures in a room

Source: Luisa Brotas.

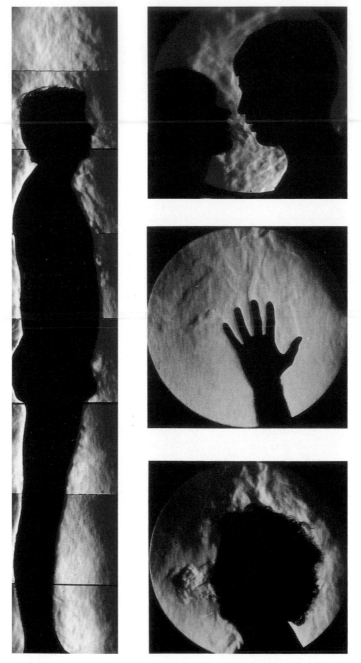

**FIGURE 2.3** A thermal image of heat plumes around people: as the heat from our skin warms the air around us it rises, driven by the buoyancy of warm air

Source: Clark and Edholm, 1985.

**FIGURE 3.5** Adaptive behaviour is not confined to humans

Source: Susan Roaf.

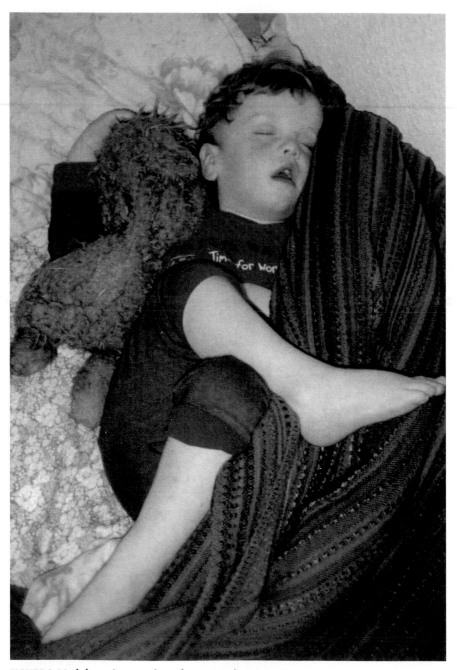

**FIGURE 3.11** Adaptation need not be a conscious act

Source: Ruth Roberts.

**FIGURE 5.3** The Japanese Cool Biz project encourages the use of informal clothing at work to reduce cooling energy use

Source: Environmental Business Women, Japan, http://www.herb.or.jp/index.html

**FIGURE 6.6** This house in Indonesia uses copious air movement coupled with an insulating roof to keep people comfortable in deep shade in a hot humid climate

Source: Hom Rijal.

**FIGURE 6.7** A heavyweight Haveli in India includes many pathways for air to move through the structure to enable occupants to adapt to changing conditions

Source: Jane Matthews.

**FIGURE 6.8** Traditional buildings in Seville have many layers between indoors and out, made of systems of curtains, blinds, glass, doors, shutters and balconies offering multiple options to adapt the building to provide comfortable conditions internally

Source: Susan Roaf.

## Esfahan, Iran 33N 52E 1550m

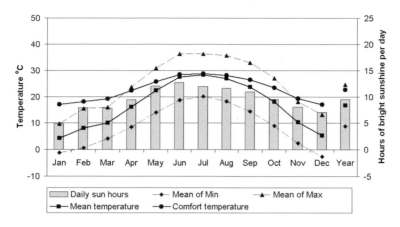

| Daily sun hours | Mean of Min | Mean of Max |
| Mean temperature | Comfort temperature | |

## Jazireh Gheshm 27N 56E 6m

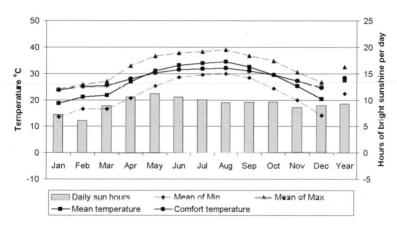

| Daily sun hours | Mean of Min | Mean of Max |
| Mean temperature | Comfort temperature | |

## Yazd 32N 54E 1230m

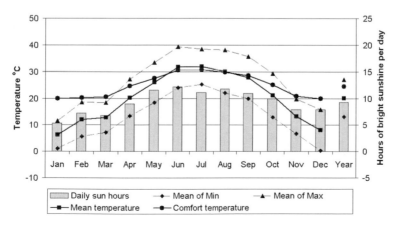

| Daily sun hours | Mean of Min | Mean of Max |
| Mean temperature | Comfort temperature | |

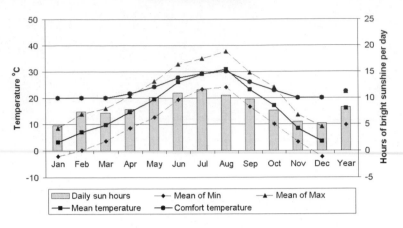

**FIGURE 6.4** The Nicol graph combines the monthly outdoor temperatures in a particular place (e.g. Esfahan) with the temperature that is predicted to be comfortable indoors. It represents the 'work' that a building must do to provide a pleasant indoor climate. This figure shows Nicol graphs for four contrasting climatic zones in Iran.

### 6.3.2 Heated or cooled buildings

Buildings that are free running for some, or even most, of the year may need to be heated or cooled in particularly cold or hot weather. In some climates suitable clothing alone is enough to ensure comfort all year round, and no heating or cooling is needed. The comfort temperature in heated or cooled buildings is a matter of custom and preference. The customary indoor temperature may be decided on local factors and may also change with the seasons as people adjust their clothing with the weather. The lowest temperature considered comfortable in Britain might be 17°C or 23°C depending on where it is, the purpose of the building, the attitudes of occupants – and who pays the bills! In Nepalese homes it might be 10K lower than that. An added complication is that a temperature indoors in the cooling months of autumn may be perceived differently in comfort terms from the same temperature in the same place in the warming months of spring. Comfort is complex!

The decision about what indoor temperature to set a control, or a thermostat, to in periods of heating or cooling will have a big effect on the energy used by the building. In the UK approximately 10 per cent of heating energy is saved for every one degree reduction in indoor temperature (see Section 1.2). It is worth making this clear to occupants, encouraging them to accept free-running operation for as long as possible, and to accept a seasonally varying indoor temperature to save energy. Thus, if the customary temperature in the heating season is 19°C (not unusual in a house in the UK) the indoor comfort temperature should not normally fall much below this temperature, even in the depths of winter, in a modern centrally heated house. In reality it often does.

In traditional buildings the method of heating a room was very different. Lower background temperatures in the building were combined with known customary 'thermal pathways' through a day. A housemaid might have been active from morning to night cleaning around a house and used her high level of metabolic heat to stay warm. Her mistress may have done nothing more energetic than write a letter or read a book but would have been sitting in front of a fire that kept her comfortable over the day. The adaptive approach to comfort enables the traditional use of local radiant heating sources with cool background temperatures to be understood as quite acceptable, and will stimulate the generation of new and efficient radiant heating appliances (Humphreys *et al.*, 2011).

### 6.3.3 Variability

The Nicol graph suggests an optimum temperature which a building designer should aim to achieve. This is not the end of the story; other things need to be considered:

- How far can we range from the optimum without causing discomfort?
- Are there limits to the comfort temperature?
- How quickly does the comfort temperature change?

We remember that change in indoor temperatures can be caused by actions taken to reduce discomfort, as well as by those that are out of the occupant's control and therefore likely to cause discomfort. Achieving thermal comfort is therefore affected by the possibilities for change, as well as the actual temperature. The width of the comfort 'zone' is highly influenced by the physical and social constraints in the building. In a situation where there is no possibility of changing clothing or activity and where air movement cannot be used, the comfort zone may be as narrow as ±2K. In situations where these adaptive opportunities are available and appropriate the comfort zone may be considerably wider.

**FIGURE 6.5** A high-backed chair and screen to protect from draughts would have helped keep the wealthy warm by the fire, while the maids did so by hard work in cleaning the house. Very different 'upstairs/downstairs' thermal pathways through the same day and house were customary

Source: Image from an 1885 catalogue in Wright, 1964.

Clearly there are limits to the range of temperatures that are acceptable. Indoor temperatures that require us to wear a thick coat will be incompatible with many indoor activities such as office work. Temperatures that require high rates of sweating may be unacceptable even though thermal balance is maintained. It would be difficult to put absolute limits on these temperatures as they would depend on local customs as well as physiology and acclimatisation.

The Nicol graph shows that people's comfort temperature changes seasonally. The way the graph is constructed (using monthly data) may suggest that the comfort temperature changes only from month to month. In fact, the change can be much quicker. Research has shown that the characteristic time taken is about a week (see Chapter 4). This means that if the weather changes from week to week, the comfort temperature will also change. Thus the Nicol graph does not say what the comfort temperature will actually be at any given time, but merely suggests its average value in that particular month of the year.

### 6.3.4 Appropriate design strategies

As we mentioned in Chapter 5, comfort temperatures obtained from field surveys in general have been rising (Humphreys *et al.*, 2010). This rise suggests that buildings may now be generally less good at keeping the heat out and better at keeping it in (see Section 6.2 above). It may also mean that the temperature in buildings is running higher because we have more heat-producing equipment inside them, and we have adapted to this higher temperature. Yet again, it may be that insufficient levels of thermal mass in modern buildings are increasing the diurnal indoor temperature swings, driving up the peak daytime temperatures. All these considerations need unravelling to understand this warming trend and attempts should be made to eliminate it, as there is only so much increased indoor temperature populations can adapt to before acceptable warmth becomes heat stress. Figures 6.6–6.8 (in Plates) illustrates the interaction between design strategy and climate in different circumstances.

There is a clear need for us to make our buildings more resilient in a warming climate with more frequent extremes, rather than less robust as all too often appears to be the case today. Buildings that are *passive, adaptive and free running* can help us remain comfortable and safe, well into the twenty-first century, with mechanical assistance as required. An adaptive approach to the design of such buildings is the subject of volume 3 of this trilogy on comfort.

### Appendix to Chapter 6

### How to make a Nicol graph

1    From the nearest met station (or library) get for each month of the year:

   - the mean daily maximum of the outdoor air temperature ($T_{omax}$);
   - the mean daily minimum of the outdoor air temperature ($T_{omin}$);
   - the mean daily solar radiation on the horizontal (optional).

   (Preferably these should be averages for several years, but use the best you can find.)

2    Determine from local people (either from experience or by conducting a survey) what the customary temperature limits are in the type of building you are designing.

3    Calculate the value of the mean outdoor temperature $T_{om}$ for each month:

$$T_{om} = T_{omax} + T_{omin})/2$$

4    Use the Humphreys formula to find the comfort temperature $T_c$:

$$T_{comf} = 0.53\ (T_{om}) + 13.8$$

5    If $T_c$ falls below customary comfort limits, for instance indicating that people will be comfortable at temperatures lower than seems reasonable, particularly in winter, then flatten off the winter comfort line at a point at which we know that people, well-clothed in a good passive building, would be comfortable.
6    Plot $T_{omax}$, $T_{omin}$, $T_{om}$ and $T_{comf}$ for each month of the year.
7    Where it is available, show the solar irradiation as a monthly bar as a guide to the possibilities for passive solar heating (the Roaf extension).

## References

Ackerman, M. (2002) *Cool Comfort: America's romance with air-conditioning*. Washington, DC: Smithsonian Institution Press.

Aynsley, R. (2012) Large ceiling fans. *Architectural Science Review* 55(1), 15–25.

Baird, G. (2010) *Performance in Practice: Mixed mode, passive and environmentally sustainable buildings – designer and user perspectives*. London: Routledge.

Black, F. and Milroy, E. (1966) Experience of air-conditioning in offices. *Journal of the Institute of Heating and Ventilating Engineers* September, 188–196.

Bryan, H. (2011) Personal communication from Professor Harvey Bryan, University of Arizona.

Cândido, C., de Dear, R.J., Lamberts, R. and Bittencourt, C. (2010) Cooling exposure in hot humid climates: Are occupants 'addicted'? In S. Roaf (ed.) *Transforming Markets in the Built Environment: Adapting for climate change*. Oxford: Earthscan.

CEN (2007) Standard EN15251. *Indoor Environmental Parameters for Design and Assessment of Energy Performance of Buildings: Addressing indoor air quality, thermal environment, lighting and acoustics*. Brussels: Comité Européen de Normalisation.

DOE (2011) Building energy databook. Available at http://buildingdatabook.even.doe.gov (September 2011).

Givoni, B. (1991) Performance and applicability of passive and low-energy cooling systems. *Energy and Buildings* 17, 177–199.

Haves, P., Roaf, S. and Orr, J. (1998) Climate change and passive cooling in Europe. *Proceedings of PLEA Conference*. Lisbon, 463–466.

Humphreys, M.A. (1978) Outdoor temperatures and comfort indoors. *Building Research and Practice (J CIB)* 6(2), 92–105.

Humphreys, M.A. and Nicol, J.F. (1998) Understanding the adaptive approach to thermal comfort. *ASHRAE Transactions* 104(1), 991–1004.

Humphreys, M., Nicol, F. and Roaf, S. (2011) *Keeping Warm in a Cool House*. Edinburgh Historic Scotland Technical Paper 14.

Humphreys, M.A., Rijal, H.B. and Nicol, J.F. (2010) Examining and developing the adaptive relation between climate and thermal comfort indoors. *Proceedings of Conference on Adapting to Change: New Thinking on Comfort, Cumberland Lodge, Windsor, UK, 9–11 April 2010*. London: Network for Comfort and Energy Use in Buildings. Available at http://nceub.org.uk.

Klinenberg, E. (2002) *Heatwave: Social autopsy of disaster in Chicago*. Chicago, IL: University of Chicago Press.

Loudon, A.G. (1968) *Window Design Criteria to Avoid Overheating by Excessive Solar Gains*. BRS Current Paper 4/68, Garston, Building Research Station.

Loudon, A.G. and Danter, E. (1965) Investigations of summer overheating. *Building Science* 1, 89-94.

Loudon, A.G and Keighley, E.C. (1964) User research in office design. *Architect's Journal* 139(6), 333–339.

McCartney, K.J. and Nicol, J.F. (2002) Developing an adaptive control algorithm for Europe: Results of the SCATs project. *Energy and Buildings* 34(6), 623–635.

Nicol, J.F. and Humphreys, M.A. (2002) Adaptive thermal comfort and sustainable thermal standards for buildings. *Energy and Buildings* 34(6), 563–572.

Nicol, J.F., Jamy, G.N., Sykes, O., Humphreys, M.A., Roaf, S.C. and Hancock, M. (1994) *A Survey of Thermal Comfort in Pakistan toward New Indoor Temperature Standards*. Oxford: Oxford Brookes University.

Rijal, H., Tuohy, P., Humphreys, M.A., Nicol, J.F. and Samuel, A. (2012) Considering the impact of situation-specific motivations and constraints in the design of naturally ventilated and hybrid buildings. *Architectural Science Review* 55(1), 35–48.

Roaf, S., Crichton, C. and Nicol, F. (2009) *Adapting Buildings and Cities for Climate Change*. Oxford: Architectural Press.

Roaf, S., Fuentes, M. and Thomas, S. (2011) *The Ecohouse Design Handbook*. London. Architectural Press.

Shove, E. (2003) *Comfort Cleanliness and Convenience*. Oxford: Berg Publishers.

Tuohy, P. (2008) Air-conditioning: The impact of UK regulations, the risks of unnecessary air-conditioning and a capability index for non-air conditioned naturally ventilated buildings. *Proceedings of the 2008 Windsor Conference on Air Conditioning and the Low Carbon Cooling Challenge*. The full paper is available at www.nceub.org.uk.

Usable Buildings Trust. Available at www.useablebuildings.co.uk (September 2011).

Ward, J. K., Wall, J. and Perfumo, C. (2012) Environmentally active buildings: the controls challenge. *Architectural Science Review* 55(1), 26–34.

Wright, L. (1964) *Home Fires Burning*. London: Routledge and Kegan Paul.

# PART II

# Practice

Conducting a survey in the field
and analysing the results

# 7

# WHAT SORT OF SURVEY?

## 7.1 Introduction

The earlier chapters have shown that field surveys are key to understanding people's interaction with their environment, and hence for the development of the adaptive approach to comfort.

The second part of the book outlines how to conduct a field survey and analyse its results. Much progress has been made in recent years not only in the theory and application of the adaptive approach to thermal comfort, but also in the development of advanced techniques of monitoring buildings and understanding the responses of their occupants. To give a detailed listing and critique of this developing technology is not possible in this small book. What we do hope to do is to give an understanding of the important variables involved, how they can be measured or gathered and, having collected data, the best ways to analyse it.

The first thing that needs to be said to anyone contemplating conducting a field survey is that you should have a clear idea of what you hope to learn from your research. This will decide what things you need to measure and how you should measure them. There is always a temptation to measure everything you can think of and then use statistics to sort it out afterwards. This temptation is increased by the ability computers have to reduce the tedium of statistical analysis. There is some reason for taking this approach – you can never be sure exactly what will be interesting – but such a scatter-gun approach to the analysis can increase the likelihood of a spurious or trivial result. If you are unclear about the meaning of the measurements you have made, you probably cannot be clear about the meaning of your results. In addition, you should always remember you are using human subjects and you don't want to ask too much from them. Chapters 8 and 10 seek to clarify these points and to make recommendations to help you build your survey method.

The second consideration is the group of people you are using as subjects. The design of your experiment must make the most of the opportunity you have to obtain information from them, and about them. Organising a comfort survey is costly in both time and effort, not only for the researcher but also and especially for your subject population. So take care to make the most of the opportunity you have been given. Chapter 9 is about experimental design.

The third consideration is the method you intend to use to draw conclusions from your survey. Again, because of the power of computers, especially in conjunction with automatic dataloggers that can now download information directly into a database, there is a temptation to move directly into sophisticated statistical techniques. These are clearly key tools that can be used, but the meaning of the results you obtain from them is not necessarily crystal clear. The emphasis in Chapter 10 is therefore on the need to be familiar with the data that your survey has yielded. From this will flow the information you can expect your data to provide, and the best method by which this can be extracted.

The adaptive approach has been criticised for being like a 'black box' that gives an end product in terms of a comfort temperature, but fails to fully explain the underlying behavioural and physical causes of comfort. The scope here for fruitful research seems almost limitless. There is a need to organise your effort so that work is not duplicated unnecessarily, and so that your results can be successfully consolidated into a consistent empirical model. A careful reading of relevant papers in journals and from conferences can help you focus on what is already known and what would be the most fruitful direction for your research.

There are lots of areas of study that would be useful to the development of the adaptive model, and we list a few below:

- How do people use thermal controls such as windows and blinds?
- Do people use subjective scales in a consistent way?
- Metabolic rate: is the present estimate by task appropriate in different climates?
- To what extent can values of clothing insulation explain the variations of the comfort temperature?
- How important are second order physical parameters such as turbulence, directional radiation, etc.?

## 7.2 The complexity of your survey

Useful information can be derived from a wide range of surveys in the field, from a single temperature measurement in an occupied room to a wide-ranging and detailed investigation involving many subjects and measurements. The level you choose to use will be decided by a number of considerations: your budget, both in time and money; what you want to find out; and how accurate your result needs to be. For convenience, we have divided field surveys into three 'levels'.

Whichever one of these levels you use for your survey, it is important that the environment that is measured is representative of the normal experience of the human subjects. This means that the conditions should be measured in a representative place in the space and during normal occupied hours of the day. Given the nature of adaptation it is also preferable that inhabitants of the space should be in a familiar environment, and allowed to respond to it as they normally would.

### 7.2.1 Level I: Simple measurements of temperature in occupied spaces, with no subjective responses

A number of such surveys have been done (e.g. Hunt and Gidman, 1981). The English House Condition Survey, whose most important consideration was the material state of the building fabric in houses, did for a number of years measure the temperature in the hallway (DETR, 1996) of each house visited. This was an important source of information about the temperatures that normally occurred in domestic settings. The TARBASE (Jenkins, 2010) and CaRB projects are more recent examples that included monitoring of domestic and non-domestic buildings. Humphreys (1995) has demonstrated just how much insight can be gained from such simple measurements.

A vital consideration with this simple method, as with any field survey, is to make sure the temperature you measure is representative. There are a number of ways in which the temperature can be taken:

a)  A spot temperature at a given time of day in a given room
b)  The temperature of a room during occupied hours
c)  The temperature of a room over 24 hours or more
d)  An average daily temperature experienced by a single person or a group.

Each of these measures gives different information:

a)  can be useful if the time of day is representative of a period of satisfaction, for example in a living room during the evening ($T_E$ in Hunt and Steele, 1980) when the occupants can be assumed to have controlled the temperature to a desirable value.

b) and c)  can also give useful information, especially if it can be assumed that the room temperature is under the control of its occupants.

d)  is, for our purposes, the most satisfactory since the neutral or comfort temperature is closely related to the average temperature experienced (see Chapter 3), and so it can be inferred from the findings.

The temperature measurements should ideally be augmented by descriptive material about the occupants and the building, and by an evaluation by the occupants of how successful the building is thermally. The temperature recorded would preferably take radiant as well as air temperature into account, and we recommend the use of a globe thermometer in such surveys (see Chapter 8).

The value of the level I survey is that it provides data about occupied spaces without much involvement from the occupants. By inference from adaptive comfort theory, the mean temperature can also give a first approximation to the indoor conditions that inhabitants of the space will find comfortable (see Figure 3.1).

### 7.2.2 Level II: Measurements of the thermal environment and the occupant response to it.

This type of survey is the classic field survey of thermal comfort in which a number of subjects provide subjective responses while at the same time the environment is measured.

The environment is taken to be represented by a 'core set' of environmental measurements: air temperature, radiant temperature, air velocity and water vapour pressure. As a minimum the subjective measures taken are the comfort vote and some scale of thermal preference. Clothing insulation and metabolic rate are also useful, though difficult to assess accurately, and are not essential. These surveys should where possible also record the use the occupants make of thermal controls such as windows, blinds and fans. It is important to keep records of the time-sequence of the comfort data. The order in which people experience temperatures and the speed with which change occurs are important in deciding their response to it.

Variations on the classic survey such as Humphreys' (1973) survey of school children with time-lapse photography (see Chapter 8) are in effect a variety of the level II survey.

### 7.2.3 Level III: Surveys that include all factors needed to calculate the heat exchange between a person and the environment, together with subjective responses

Several surveys have been undertaken where the observation of comfort votes have been augmented by more detailed data on clothing and activity (Fishman and Pimbert, 1982; Griffiths, 1990; Schiller, 1990; Williamson and Coldicutt, 1991; Busch, 1990, 1992; Nicol et al., 1999; Nicol and McCartney, 2001; Raja et al., 2001). De Dear and Brager (1998) have gathered the data from many such surveys together in order to make a meta-analysis that crosses many climates and becomes a global database, similar to that of Humphreys (1975) but allowing a more detailed picture by virtue of the inclusion of more detail in the surveys. Such surveys allow for comparison between empirical results and those calculated using heat-balance comfort models, and thus the comparison of laboratory estimates of comfort vote and findings in the field, such as were reported in Chapter 4 (Figure 4.1).

There are many other ways in which the classic comfort survey can be added to in order to supply supplementary information. For instance other environmental variables that are considered to be important in defining thermal comfort are air turbulence and radiant asymmetry. So long as the 'core' information mentioned in Section 7.2.2 is included, and measured in a comparable way, the results of different surveys can be used to add to the general pool of information.

### 7.3 Post occupancy evaluation of buildings (POE)

With the rising interest in the energy use of buildings, and the ways in which buildings can adapt to a changing climate, the evaluation of buildings has become increasingly important. In particular, the performance evaluation of occupied buildings (as opposed to the simulation of buildings before construction) is seen as essential, as it becomes clear how frequently buildings do not measure up to the predictions of their designer (Leaman, 2004). A special issue of *Building Research and Information* (vol. 33(4), 2005) contains a number of papers that arose from the Windsor Conference in 2004, centred on post occupancy evaluation.

Estimating actual building energy use and related emissions are mandatory in many countries in the European Union, under the conditions of the Energy Building Directive (EPBD, 2011). This requires that all buildings display energy performance certificates showing actual energy use in the building (operational rating), or its modelled energy use (asset rating) at point of sale, lease or rent, depending on the country and the interpretation of the EPBD they have adopted. The display of the energy performance of buildings inevitably prompts some building owners to act to improve that performance and the market rental value of their property. The best way to develop a cost-efficient action plan is to know where energy is being wasted and the best way to find this out is with a POE.

Post occupancy evaluation of buildings (Baird *et al.*, 1996) is concerned centrally with the building, unlike the comfort survey which focuses on the building occupants (Nicol and Roaf, 2005). The POE includes three elements:

- A user survey in which the inhabitants of the building are asked for their evaluation of it using questions such as 'how often does it overheat in summer?'. The adequacy of the building is being evaluated in terms of the impressions of the inhabitants. Inhabitants are in effect being used as the memory of the building.
- The physical form of the building is measured and the level of occupant control is estimated together with any special features or technologies in the building.
- An energy audit is undertaken of the building in use.

The POE, called building performance evaluation (BPE) in the USA (Preiser and Vischer, 2005), allows the researcher to identify problems and locate them in the building. It also suggests possible solutions. It also enables the research team to measure the performance of the building against other similar buildings and their performance benchmarks. More on the post occupancy evaluation of buildings is available on the website of the Usable Buildings Trust (www.usablebuildings.co.uk).

## References

Baird, G., Gray, J., Isaacs, N., Kernohan, D. and McIndoe, G. (1996) *Building Evaluation Techniques*. New York: McGraw Hill.

Busch, J. (1990) Thermal responses to the Thai office environment. *ASHRAE Trans* 96(1), 859–872.

Busch, J. (1992) A tale of two populations: Thermal comfort in air-conditioned and naturally ventilated offices in Thailand. *Energy and Buildings* 18, 235–249.

de Dear, R.J. and Brager, G.S. (1998) Developing an adaptive model of thermal comfort and preference. *Field Studies of Thermal Comfort and Adaptation*. ASHRAE Technical Data Bulletin 14(1), 27–49.

DETR (1996) *1991 English House Condition Survey: Energy Report*. London: The Stationery Office.

EPBD (2011) Implementing the energy performance of buildings directive – featuring country reports, 2010, Brussels. Available at www.epbd-ca.org (September 2011).

Fishman, D.S. and Pimbert, S.L. (1982) The thermal environment in offices. *Energy & Buildings* 5(2), 109–116.

Griffiths, I. (1990) Thermal comfort studies in buildings with passive solar features: Field studies. *Report to the Commission of the European Community, ENS35 090 UK*.

Humphreys, M.A. (1973) Classroom temperature, clothing and thermal comfort – a study of secondary school children in summertime. *J. Inst. Heat. & Vent. Eng.* 41, 191–202.

Humphreys, M.A. (1975) Field studies of thermal comfort compared and applied: Physiological requirements on the microclimate, Prague. Reprinted (1976) *J. Inst. Heat. & Vent. Eng.* 44, 5–27.

Humphreys, M.A. (1995) Thermal comfort temperatures and the habits of Hobbits. In F. Nicol, M. Humphreys, O. Sykes and S. Roaf (eds) *Standards for Thermal Comfort*. London: Spon (Chapman & Hall), 3–13.

Hunt, D. and Gidman, M. (1981) A national field survey of house temperatures. *Building and Environment* 17(2), 107–124.

Hunt, D.R.G. and Steele, M.R. (1980) Domestic temperature trends. *Heating and Ventilating Engineer* 54(626), 5–15.

Jenkins, D.P. (2010) The value of implementing carbon-saving measures in fuel-poor social housing. *Energy policy* 38, 832–839.

Leaman, A. (2004) Post Occupancy Evaluation. In S. Roaf (ed.) *Closing the Loop: Benchmarks for sustainable buildings*. London: RIBA Publications.

Nicol, F. and McCartney, K. (2001) *Final Report: (Public) Smart Controls and Thermal Comfort (SCATs)*. Oxford: Oxford Brookes University.

Nicol, F. and Roaf, S. (2005) Post occupancy evaluation and field studies of thermal comfort. *Building Research and Information* 33(4), 338–346.

Nicol, J.F., Raja, I.A., Allaudin, A. and Jamy, G.N. (1999) Climatic variations in comfort temperatures: The Pakistan projects. *Energy and Buildings* 30(3), 261–279.

Preiser, W. and Vischer, J. (2005) *Post Occupancy Evaluation*. New York: Harper Collins.

Raja, I.A., Nicol, J.F. and McCartney, K.J. (2001) The significance of controls for achieving thermal comfort in naturally ventilated buildings. *Energy and Buildings* 33, 235–244.

Schiller, G. (1990) A comparison of measured and predicted comfort in office buildings. *ASHRAE trans* 96(1), 609–622.

Williamson, T.J. and Coldicutt, S. (1991) Aspects of thermal preferences in housing in a hot humid climate with particular reference to Darwin, Australia. *Int J Biometeorol* 34, 251–258.

# 8

# INSTRUMENTS AND QUESTIONNAIRES

Four basic types of measurements are common in comfort surveys:

- Physical measurements – temperatures, humidity and air movement
- 'Personal variables' – clothing insulation and metabolic rate
- Subjective measures – thermal comfort, thermal preference, self-assessed productivity, overall comfort and so on
- Behaviour – records of adaptive measures: windows open/closed fans on/off, etc.

## 8.1 Physical measures

Thermal comfort theory (Chapter 2) suggests that there are four important physical variables:

- Air temperature $(T_a)$ °C
- Radiant temperature $(T_r)$ °C
- Air velocity (v) m/s
- Water vapour pressure $(P_a)$ $kP_a$.

There are also the 'personal variables':

- Clothing insulation (clo)
- Metabolic rate (met).

In this chapter we shall look at suitable ways to make these measurements.

There are standard protocols for collecting physical data for the estimation of thermal comfort. Both ISO and CEN standards refer to ISO 7726 (2001) which gives details of how to make precise measurements. The precision required by this protocol is not necessary in many field surveys and also requires a generous budget. Researchers from the Politecnico di Milano who were using the requisite instrumentation to test EN15251 report that the equipment is very costly (Zangheri, 2010).

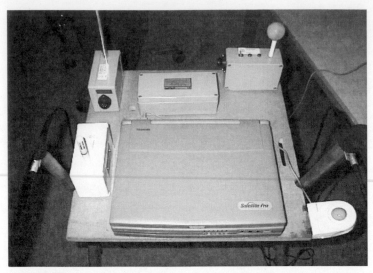

**FIGURE 8.1A** The SCATs instrument set (McCartney and Nicol, 2002) which included measurements of the environmental variables using a range of sensors and the collection of subjective measures by recording of answers to a number of questions

Source: John Stoops.

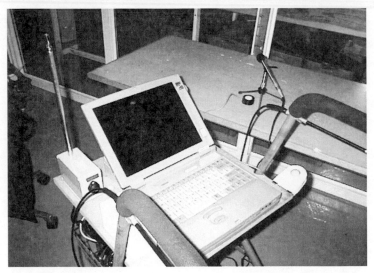

**FIGURE 8.1B** The SCATs instrument set shown on the trolley that was used to transport it from room to room and from building to building. The instruments were allowed to 'settle' for some minutes before their readings were polled and the subject was questioned for their subjective responses

Source: John Stoops.

### 8.1.1 Air temperature $T_a$ (°C)

Any object will respond to both the convective and the radiant environment: it will exchange heat with the air by convection and with the surrounding surfaces by radiation. So the final temperature of the object is the temperature at which the total heat exchange is zero. Thus, if the radiant temperature is different from the air temperature (and we have to assume that it is) then any temperature probe will be at some temperature intermediate between the radiant and air temperatures.

So, to measure the air temperature we need to minimise the effect of heat radiation. In early field studies air temperature was usually measured using a whirling or sling hygrometer (Figure 8.2). This is an instrument with two thermometers in a wooden frame. One thermometer has its bulb enclosed in a cotton wick that is moistened with distilled water. The purpose of this is to obtain the 'wet bulb temperature' which allows the researcher to calculate the water vapour pressure (of which more later). The other thermometer has its bulb open to the surrounding air. The purpose of the sling is to allow the thermometer to be 'whirled' in the air. This increases convective heat exchange with the air, making the radiant effects relatively less important. Whirling hygrometers are rarely used nowadays and are obtrusive and demanding in use.

Another method of reducing the effect of radiation is to place a bright metallic shield around (but not touching) the thermometer's bulb or sensor. Metals have a low emissivity, which means that their radiant exchange with their surroundings is small. The shield cuts out the direct radiant component while allowing the thermometer to maintain close contact with the air. Thermocouples and thermistors supplied with dataloggers are sometimes embedded in a metallic sheath, which reduces the radiant component and shields the sensor – but it increases the instrument's size and its thermal inertia. Another way to cut down the relative effect of radiation is to reduce the size of the sensor. The smaller the sensor, the greater the convective exchange becomes per unit surface area, while the radiant exchange per unit area remains unchanged (Figure 8.3).

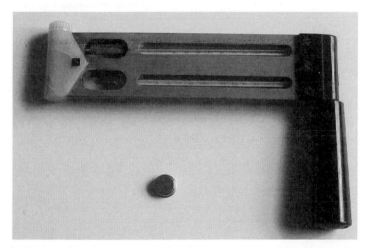

**FIGURE 8.2** Whirling or sling hygrometer. The thermometers are 'whirled' in the air to reduce the effect of radiant heat. A miniature i-Button datalogger is shown for comparison

Source: F. Nicol.

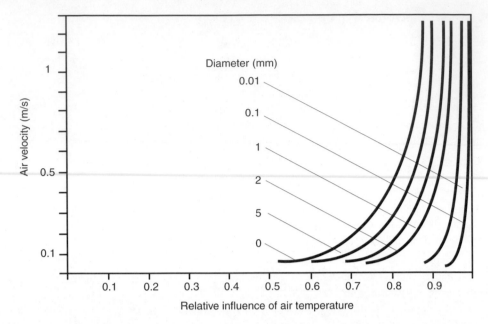

**FIGURE 8.3** Relative influence of air temperature and mean radiant temperature on the temperature measured by an instrument for different air velocities and sensor dimensions. Note that the smaller the instrument and the higher the air velocity the nearer its temperature is to the air temperature

Source: BS ISO 7726.

These days it is common to use dataloggers that collect and store data at preset intervals over a period of time. The loggers will often use thermocouples or thermistors as temperature sensors. They have the advantage of cheapness and availability, and can give accurate results. Unshielded thermocouples are usually sufficiently small to be insensitive to low-temperature radiation from indoor surfaces, though the direct sun should not be allowed to fall on them. Thermistors, despite their slightly greater size, can be equally good, and logging equipment is often designed to use them. In all cases the reading will become closer to the air temperature if the thermometer head is ventilated with (unheated) air.

Small stand-alone dataloggers with built-in sensors are increasingly common and claim to measure air temperature (and in some cases relative humidity too). You will need to satisfy yourself that these loggers are indeed measuring air temperature as their size can mean that they are also sensitive to thermal radiation. Many are comparable in size to a 40mm globe thermometer (see below) and may record a temperature that is close to the globe temperature.

Miniature temperature loggers have been developed by the food industry and are comparable in size to a small button-cell (10 to 15mm diameter) (Figure 8.2). They are designed to be used to check that food remains at the right temperature while being transported. These loggers have the great advantage of being very cheap to buy, but they may not discriminate changes less than 0.5 or 1K. They are not so small that they won't react to radiant temperature, so they may need to be shielded if the intention is to measure air temperature; however, some have shiny metal outer cases that will much reduce their sensitivity to radiant heat.

### 8.1.2 Globe temperature $T_g$ (°C)

The globe temperature is measured using a sphere, generally painted black or grey to mimic the response of the human body to thermal radiation, and is nowadays commonly about 40mm in diameter. The temperature is measured at the centre of the sphere. From the foregoing discussion it is clear that the globe temperature will be somewhere between air temperature and radiant temperature. The reason we use a 40mm sphere is because the balance between the two temperatures for a sphere of this size most closely resembles that for the human body (roughly half-and-half at low air speeds). The reason that such a small sphere has this property is because the human body is a complex shape, with some of its surfaces radiating to each other more than to the room surfaces, and so it reacts less to radiant temperature than a sphere of a comparable size would do. The theoretical derivation of this can be found in Humphreys (1977); 40mm is roughly the diameter of a table tennis ball.

For a simple level I survey (Chapter 7) with only one measuring instrument you should use a globe thermometer as a measure of temperature because it reacts to the thermal environment in much the same way as a human body. Some computer-modelling techniques for estimating indoor temperatures in buildings output a weighted average of surface (radiant) and air temperature sometimes called the 'environmental temperature' which, though not precisely the same as the operative temperature (see Section 8.1.4) or the globe temperature, is nearer to them than to the air temperature (Humphreys, 1974).

The globe thermometer takes some minutes to reach equilibrium. There is some merit in this in a variable environment, since it smoothes out temperature fluctuations over a period. The time constant can be determined experimentally (Humphreys, 1974).

A globe thermometer can either be purchased ready-made or be made up yourself using a black or grey-painted table-tennis ball with the temperature-measuring device inside it. The material from which the globe is made and its thickness will affect the time it takes to come to equilibrium. The response time is also affected by the properties of the temperature sensor. The sensor should be located at the centre of the sphere and the entry point of the thermometer stem or wire should be made fairly airtight, because air flowing into the globe can affect the air/radiant temperature balance. Note also the comment above, that a small datalogger may respond like a globe thermometer, and be used in its place.

### 8.1.3 Radiation measurement and mean radiant temperature $T_r$ (°C)

Thermal radiation is more complex to measure than air temperature. The radiation, being dependent on the temperatures of the several surrounding room surfaces, will vary not only with the position in the room, but also with the direction in which it is measured. This means that when fully describing a radiant environment you should include the direction in which the radiation is measured.

It is possible to build a detailed picture of the radiant environment in a room from the electronic images taken by an infra-red camera (see Figure 2.2 in Plates). The images will indicate the temperatures of the room surfaces. Alternatively you can use a surface-temperature 'gun'. This instrument can be 'aimed' at any (non-metallic) indoor surface and will display its temperature. Doing such surveys automatically would not only require costly equipment but would also require complex analysis.

International Standard ISO 7730 (2005) requires the evaluation of radiant asymmetry in defining the thermal environment. Methods are available in ISO 7726 (2001) for evaluating this asymmetry using the instruments they describe, or other kinds of radiometer. For more information on the measurement of directional radiation see, for instance, McIntyre (1980).

Such complexity is rarely necessary. It is usually sufficient to obtain the 'mean radiant temperature' (MRT). This is defined as 'the temperature of a sphere at the point in question which would exchange no net radiation with the environment'. In effect, this means the radiation is averaged for all directions and so the resulting 'mean radiant temperature' has a single value for any particular point in the room. Unless your experiment concerns the distribution of radiation from different directions then the MRT will be sufficient.

In most comfort surveys mean radiant temperature is obtained from a calculation using the air temperature, globe temperature and air speed. The spherical shape of the globe means that it will integrate the radiation in the way required for MRT estimation. A formula (see equation 8.1 below) gives the value for the MRT ($T_r$). A word of caution: because $T_r$ is estimated from the difference between $T_g$ and $T_a$ the effects of any errors in $T_a$ and $T_g$ are magnified by an amount dependent on the value of v. Any error in the measurement of $T_a$ or $T_g$ will mean that $T_r$ will have a much greater error – more than double the error in $T_a$ or $T_g$ if the air is still, and even more at higher air speeds. Because of this it is worth considering using the globe temperature itself as a variable rather than the mean radiant temperature, particularly if you have only low-precision instruments. The globe temperature has been so widely used in comfort surveys that $T_g$ has almost become a basic variable, and so long as the globe is of standard diameter, the temperatures have the same meaning from one survey to another.

One other problem needs to be mentioned here: there are other definitions of mean radiant temperature in the literature on thermal comfort. Fanger's definition (1970), among others, refers to the radiant exchange as experienced by a human-shaped body centred at the point in question, rather than by a small sphere. Others have used the term to mean the average surface temperature in a room. You would do well to look carefully at the definition of the term in any particular instance.

The mean radiant temperature is calculated from $T_g$, $T_a$ and the air velocity v for a globe of diameter d:

$$T_r = [(T_g + 273)^4 + (1.2 \times 10^8 d^{-0.4})v^{0.6}(T_g - T_a)]^{0.25} - 273 \qquad (8.1)$$

For a 40mm (0.04m) diameter globe this approximates to

$$T_r = T_g + 4.02\sqrt{v}\,(T_g - T_a) \qquad (8.2)$$

Alternatively ISO 7726 provides the nomogram in Figure 8.4 from which a factor $\Theta$ for the balance between air and radiant temperature can be read for different air velocities and globe diameters:

$$T_g = \Theta T_a + (1 - \Theta)T_r \qquad (8.3)$$

The globe temperature method is usually used for the estimation of $T_r$ because it is both cheap and relatively simple. If you need an accurate measurement of $T_r$ then a more sophisticated instrument is needed. But if the globe temperature is your only option do ensure that the measurements you make of air temperature, globe temperature and air velocity are as accurate as possible. The accurate measurement of air velocity is particularly difficult.

### 8.1.4 Operative temperature ($T_{op}$) °C

The operative temperature ($T_{op}$) is an index that combines the air temperature and the mean radiant temperature into a single value to express their joint effect. It is a weighted average of the two, the weights depending on the heat transfer coefficients by convection ($h_c$) and by radiation ($h_r$) (see Chapter 2) at the clothed surface of the occupant. It is often used to express the temperature of a space. Thus in both the ASHRAE 55 and the CEN EN15251 standard the comfort temperature in the adaptive section is expressed in terms of the operative temperature (see Figures 5.1 and 5.2).

The operative temperature is defined in Section 2.5. At indoor air speeds below 0.1 m·s⁻¹, it is approximated by the equation

$$T_{op} \approx \tfrac{1}{2}\,T_a + \tfrac{1}{2}\,T_r \qquad (8.4)$$

Operative temperature is a theoretical and not an empirical measure and therefore cannot be measured directly, but in practice it approximates very closely to the 40mm globe temperature .

In well-insulated buildings and away from direct radiation from the sun or from other high temperature radiant sources, the difference between the air and the mean radiant temperature (and hence between the air, the globe and the operative temperatures) is small. But notice that this does not apply outdoors.

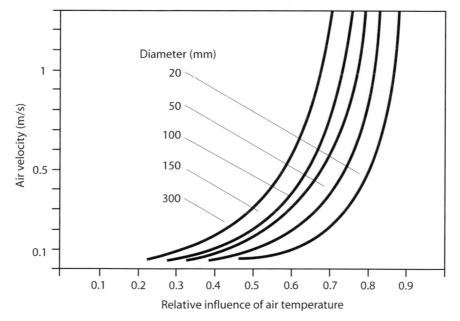

**FIGURE 8.4** Relative influence of air temperature and mean radiant temperature on the globe temperature for different air velocities and globe diameters

Source: BS ISO 7726.

### 8.1.5 Air velocity v (m/s)

To start with, it is necessary to consider how important air movement is likely to be in the circumstances you are measuring. In a closed room in a temperate climate, air movement is likely to be slight, particularly away from the walls. Such low velocities are very difficult to measure. Nevertheless, air currents can occur near cold or hot surfaces or near ventilation inlets, and some measurements of these are worth making.

You should remember that convective air currents occur naturally around the human body (see Figure 2.1). Even from a seated person, air heated by the body rises in a 'plume' above the head in otherwise still air conditions. So in these common conditions the only noticeable air movement may be that in the convection plume around and above the person. For this reason the accurate measurement of air velocities below 0.1m/s is unnecessary and could be misleading.

It is worth starting by using visualisation techniques to investigate the pattern of air movement in the spaces occupied by your subjects. This can be done in a number of ways. Most convenient is to use some thistledown, or small helium-filled balloons weighted so that they neither rise nor fall in still air. Other more complex methods are available but may not be practical. If you can, move around the room and locate draughts, or the limits of an air jet. Build up a picture of the pattern of air movement in the room and think about the effect this might have on your subjects.

In naturally ventilated spaces, especially in a hot climate, air movement will have a big effect on the thermal comfort of the subject. Nicol (1974) in a field study and Rohles and Nevins (1971) in the laboratory both estimate the effect of air movement to be the equivalent of a drop in temperature of about three degrees. The reasons for the change are twofold: first, air movement evaporates sweat, making the skin dryer (reducing the 'skin-wettedness'); and second, the heat exchange by convection will be increased. (A seated person in light clothing will be comfortable at a temperature of 30°C if there is some air movement.)

The measurement of air velocity in a room presents a number of problems, principally because it is so variable from place to place and time to time, both in direction and magnitude. The problem for the comfort researcher is to represent all this variability in a meaningful way, for while it may be possible to measure all the characteristics of the air velocity, it would be difficult to know what to do with the data when you had it.

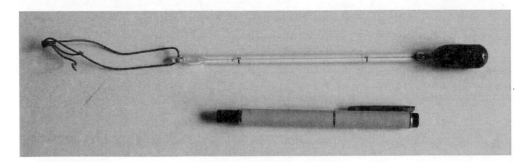

**FIGURE 8.5** The Kata thermometer, used in many early surveys for estimating the air velocity in a space

Source: F. Nicol.

The traditional way to measure air speed in thermal comfort studies was to use a Kata thermometer. This is an alcohol-in-glass thermometer with a large silvered bulb (Figure 8.5). The stem of the thermometer is marked with only two graduations, 3K apart. The method is to heat the thermometer to a temperature above the upper graduation and then to measure the time the alcohol column takes to drop between the marks, and to use this cooling time to estimate the air speed. The air speed measured by the Kata thermometer integrates the combined effect of turbulence and velocity over a substantial period of time. Using a Kata thermometer was time consuming and demanding, requiring patience and persistence. Nowadays they are expensive and hard to obtain.

Some other methods of measuring air velocity also rely on the effect of air movement on a heated body. Commonly used types are hot-wire and heated-sphere anemometers.

The hot-wire anemometer consists of a fine wire suspended between two supports and through which an electrical current is passed. There are two types: constant current or constant temperature. In the former the temperature of the hot wire is the measured variable, while in the latter it is the current needed to maintain the wire at a constant elevated temperature. The constant temperature form has advantages in reproducibility, but requires more sophisticated control circuitry. The hot-wire anemometer has advantages in speed of reaction, but there is a problem of directionality, since flow across the wire will produce a different cooling effect from flow up or down its length. Relatively cheap hot-wire anemometers are available but their response can be very directional, having been developed to measure air flow in ducts, where the direction of flow is known.

The heated-sphere anemometer, in which the heated element is embedded in a spherical head, partially overcomes the problem of directionality, certainly in the horizontal plane. The size of the sphere is crucial, both for the reaction time of the anemometer and for its interaction with the thermal environment. The effect of thermal radiation discussed above in the section on $T_a$ will apply equally to the sensor of an anemometer. The best solution is to keep the sphere small, perhaps similar in size to an air temperature sensor. Since the air velocity is calculated from the difference between air temperature and the temperature of the heated element, and both will be affected to much the same extent by thermal radiation, such a precaution is probably sufficient.

Again there are problems with a small probe. Its very ability to react quickly means that any spot reading may not be typical of the usual air movement in the room. The air velocity relevant to human comfort is some combination of average velocity over a period of time and the fluctuations in instantaneous velocity caused by turbulence. Fanger et al. (1988) demonstrated a separate effect of air turbulence on the thermal sensation. A full description of the cooling caused by elevated air movement takes account of turbulence as well as the time-averaged air movement.

When averaging the air velocities, it might seem logical to take a simple average of velocity over a set time period, say three minutes. But remember that we are interested in the air velocity for its cooling power and not its absolute value. The cooling power is more closely described by the square root of the air velocity (see Chapter 2), so averaging its square root ($\sqrt{v}$) will give a more useful result.

A well-tried anemometer is the rotating-vane anemometer, much like a child's pinwheel but much more carefully engineered. This is easy to use but of very limited accuracy at low air velocities. It can be useful where fans are in regular use and air velocities are relatively high. For outdoor wind measurements the cup anemometer is most often used because a robust instrument is necessary.

The absence of air velocity as a measured parameter will reduce the credibility of your results. Your choice of anemometer will be constrained by the kind of conditions you are measuring and the size of your budget. Of the various small-body anemometers generally available for use with dataloggers the heated bead has the advantage of being relatively non-directional and the constant-temperature form has the advantage mentioned over the constant current. All heated anemometers have the problem that a heating current is needed, and this is a drain on power supplies. As a result, many dataloggers come without an anemometer and you may need to rely on spot readings or develop alternatives. For instance where fans are available one might develop a relationship between the fan setting and the air movement, and then simply keep a record of the fan settings.

### 8.1.6 Water vapour pressure $P_a$ (kP$_a$)

In the majority of surveys the humidity of the air has little effect on thermal comfort. It is an important variable to measure only in warm and hot conditions. Most researchers give their results in terms of the relative humidity, but we advise the use of the water vapour pressure as the index of humidity, rather than the relative humidity or wet bulb temperature. This is because it is the most basic measure of the water in the atmosphere, and the most relevant measure for the physics of heat exchange (see Chapter 2). The water vapour pressure is that part of the total atmospheric pressure that is exerted by the water vapour in the atmosphere. The relative humidity is the ratio, usually expressed as a percentage, of this to the maximum water vapour pressure possible at the temperature in question (the amount of water the air can contain varies with temperature). The rate at which the air absorbs water from the body by evaporation is decided by the vapour pressure at skin temperature in relation to vapour pressure in the surrounding room air. The relative humidity at air temperature is of less relevance. Many instruments register relative humidity, but this can easily be converted to $P_a$ if $T_a$ is known. $P_a$ can be estimated from the relative humidity (RH) and the air temperature $T_a$ using the formulae:

$$P_{as} = 0.611\exp(17.27T_a/(T_a + 273)) \tag{8.5}$$

$$P_a = P_{as}RH/100 \tag{8.6}$$

where $P_{as}$ is the saturated water vapour pressure and the RH is expressed as a percentage.

The traditional manual method of finding $P_a$ was by measuring the air temperature and the wet bulb temperature using a whirling hygrometer (see above) and calculating $P_a$ from the difference between them (or reading it off a hygrometric chart). As with many other traditional methods, this has the advantage of being a well-refined and documented method that is reasonably accurate, but is tedious and time consuming (the instrument should be whirled for four minutes to ensure that the wet-bulb temperature has reached its 'true' value). The wick must be clean from dissolved salts, and distilled water, not tap water, should be used. An alternative wet and dry bulb hygrometer is the Assman Psychrometer, which uses a clockwork fan to draw air over the thermometers. This gives more reproducible results, since the clockwork fan does not get tired of whirling as the experimenter might do, and the thermometer bulbs are placed within radiation shields, but it is a bulky instrument and not very easy to use in a field study. The use of a wet- and dry-bulb thermocouple is possible with a datalogger, but the value of the wet-bulb temperature will be difficult to interpret unless a fan is used to blow air over the thermocouples. Another danger with all wet-bulb instruments is that the distilled water in the reservoir can become contaminated or that the wick will become dry, so giving misleading results.

The most accurate method of finding $P_a$ is by using a dew-point hygrometer. It measures the temperature at which the water from the air begins to condense on a mirror (the dew point). This temperature enables the water vapour pressure to be calculated from equation 8.5 by replacing the air temperature with the dew-point temperature. Modern instruments are electronically controlled and automated – and expensive.

Most humidity probes for dataloggers depend on the change of the capacitance of a semi-conductor at different humidities, and typically quote accuracies of between 2 and 5 per cent on the RH scale, at relative humidities up to 80 per cent. This is adequate for most purposes except perhaps in hot humid climates. Some manufacturers quote a range of 0–100 per cent RH for their sensors, but you should treat such claims with caution, especially if you are working in a hot climate, as humidity is an important measure in such conditions. An occasional check against a whirling or Assman hygrometer is wise, as the calibration of semi-conductor capacitance sensors can drift over the months.

We suggest, if you are using a datalogger, you use a capacitance probe for measuring relative humidity unless humidity is particularly important in your experiment. Humidity probes are expensive relative to temperature probes, but it is possible to limit the number of places at which humidity is measured, since $P_a$ varies little from place to place in a room. Surveys not using a datalogger can use an Assman or a whirling hygrometer to measure the humidity.

### 8.1.7 Accuracy of physical measurements

There is a common confusion between the accuracy of a measurement and the apparent precision of the instrument. It is not uncommon to have a digital display that shows a temperature to the nearest 0.1° when the thermometer is only accurate to the nearest 0.5° or even 1°. When choosing your instruments you should ensure that the accuracy of the measurement (not the apparent accuracy of the display) is sufficient for your purposes. The accuracy you should aim for is:

- Temperatures: ±0.5K for general temperature measurements. This accuracy should be achievable with thermocouples, cheap thermistors and many cheap dataloggers. As we discussed, if the globe temperature is to be used to evaluate MRT, then the accuracy of the globe and air temperature measurements (as well as of the air velocity) will need to be high and an accuracy of ±0.2K should be the aim. This will probably require the use of thermistors and the two thermistors (air and globe) should be matched.
- Air velocity: an accuracy of 10 per cent should be aimed at, but this may be hard to achieve at low velocities.
- Water vapour pressure: many humidity probes are calibrated in per cent RH; 5 per cent RH seems a reasonable accuracy.

It is wise to check the accuracy of the instruments before and after the experimental period by comparison with instruments of known accuracy. Ideally a full calibration should be obtained.

## 8.2 Personal variables

### 8.2.1 Clothing (clo)

Knowledge of the insulation of the clothing is not necessary for an estimate of the comfort temperature in a given situation. However, the clothing is a function of the climatic and social milieu of the subjects and therefore one of the factors that decide the desired conditions, so it is good to record it. If the results of your survey are to be used to compare observed comfort vote and PMV, then knowledge of the clothing insulation is essential. There is unfortunately no satisfactory way to measure the clothing insulation in a survey, so recourse is made to estimation.

McCullough and her co-workers (1985) produced a comprehensive list of clothing descriptions that can be used to estimate clothing insulation. The result is in clo-units (one clo-unit is approximately the insulation of a suit with normal underclothes and equals $0.155$ K.m$^2$/W). The clo is the unit often used in comfort equations, but be alert, as some programs require the input to be in K.m$^2$/W. Some examples of the insulation of particular clothing ensembles are given in Table 8.1 and more complete tables including the measured contribution of particular garments are available in ISO Standard 9920 (2009) which now includes a selection of non-Western clothing ensembles (see Figure 2.4).

### 8.2.2 Metabolic rate (met)

The measurement of metabolic rate is also not necessary in field studies, being a function of the social and climatic milieu and of the task of the group of people for which a comfort temperature is being found. Nevertheless, if the results of your survey are to be used to compare observed comfort vote and predicted PMV, then knowledge of the metabolic rate is essential. There is no satisfactory way to measure it in a survey, so recourse is again made to estimation.

A standard list of the metabolic rates from ISO 8996 (2004) is given in Table 8.2. Metabolic rate is often measured in met units where 1 met = $58.2$ w/m$^2$ (about 100 watts altogether for a 'standard' person) and is the heat given out by a seated person. Notice that the metabolic rates are assumed to be determined solely by the task being undertaken by the subject. It is also worth remembering that the measurements of metabolic rate upon which these tables are based were made in steady-state conditions, as were the determinations of their effect on comfort in the heat balance model. It would be best to include a question about activity in your comfort questionnaire to help you to determine the best available description of the activity of your subject each time you get a comfort vote.

A complication that may arise is that people can vary their activity from time to time, and a particular activity can affect the thermal sensation for some time afterwards. One way of dealing with this is to ask for a proportional breakdown of different activities of the last half hour or hour, so that a time-weighted mean can be calculated. Such matters need to be discussed with the subjects beforehand, and some method of resolving the problem worked out. If a proportional breakdown of activities is not practical, a simple note that the activity has

**TABLE 8.1** Some examples of the insulation of particular clothing ensembles (from ISO Standard 9920 2009, reproduced with permission)

| Working clothing | clo | $I_d$ $m^2.K.W^{-1}$ | Daily wear clothing | clo | $I_d$ $m^2.K.W^{-1}$ |
|---|---|---|---|---|---|
| Underpants, boiler suit, socks, shoes | 0.7 | 0.11 | Panties, T–shirt, shorts, light socks, sandals | 0.3 | 0.05 |
| Underpants, shirt, trousers, socks, shoes | 0.75 | 0.115 | Panties, petticoat, stockings, light dress with sleeves, sandals | 0.45 | 0.07 |
| Underpants, shirt, boiler suit, socks, shoes | 0.8 | 0.125 | Underpants, shirt with short sleeves, light trousers, light socks, shoes | 0.5 | 0.08 |
| Underpants, shirt, trousers, jacket, socks, shoes | 0.85 | 0.135 | Panties, stockings, shirt with short sleeves, skirt, sandals | 0.55 | 0.085 |
| Underpants, shirts, trousers, smock, socks, shoes | 0.9 | 0.14 | Underpants, shirt, lightweight trousers, socks, shoes | 0.6 | 0.095 |
| Underwear with short sleeves and legs, shirt, trousers, jacket, socks, shoes | 1 | 0.155 | Panties, petticoat, stockings, dress, shoes | 0.7 | 0.105 |
| Underwear with short legs and sleeves, shirt, trousers, boiler suit, socks, shoes | 1.1 | 0.17 | Underwear, shirt, trousers, socks, shoes | 0.7 | 0.11 |
| Underwear with long legs and sleeves, thermojacket, trousers, socks, shoes | 1.2 | 0.185 | Underwear, track suit (sweater and trousers), long socks, runners | 0.75 | 0.115 |
| Underwear with short sleeves and legs, shirt, trousers, jacket, thermojacket, socks, shoes | 1.25 | 0.19 | Panties, petticoat, shirt, skirt, thick knee socks, shoes | 0.8 | 0.12 |
| Underwear with short sleeves and legs, boiler suit, thermojacket and trousers, socks, shoes | 1.4 | 0.22 | Panties, shirt, skirt, roundneck sweater, thick knee socks, shoes | 0.9 | 0.14 |
| Underwear with short sleeves and legs, shirt, trousers, jacket, thermojacket and trousers, socks, shoes | 1.55 | 0.225 | Underpants, singlet with short sleeves, shirt, trousers, V-neck sweater, socks, shoes | 0.95 | 0.145 |

*Continued*

**TABLE 8.1** *continued*

| Working clothing | clo | $I_d$ $m^2.K.W^{-1}$ | Daily wear clothing | clo | $I_d$ $m^2.K.W^{-1}$ |
|---|---|---|---|---|---|
| Underwear with short sleeves and legs, shirt, trousers, jacket, heavy quilted outer jacket and overalls, socks, shoes | 1.85 | 0.285 | Panties, shirt, trousers, jacket, socks, shoes | 1 | 0.155 |
| Underwear with short sleeves, trousers, jacket, heavy quilted outer jacket and overalls, socks, shoes, cap, gloves | 2 | 0.31 | Panties, stockings, shirt, skirt, vest, jacket | 1 | 0.155 |
| Underwear with long sleeves and legs, thermojacket and trousers, thermojacket and trousers, socks, shoes | 2.2 | 0.34 | Panties, stockings, blouse, long skirt, jacket, shoes | 1.1 | 0.17 |
| Underwear with long sleeves and legs, thermojacket and trousers, parka with heavy quilting, overalls with heavy quilting, socks, shoes, cap, gloves | 2.55 | 0.395 | Underwear, singlet with short sleeves, shirt, trousers, jacket, socks, shoes | 1.1 | 0.17 |
| | | | Underwear, singlet with short sleeves, shirt, trousers, vest, jacket, socks, shoes | 1.15 | 0.18 |
| | | | Underwear with long sleeves and legs, shirt, trousers, V-neck sweater, jacket, socks, shoes | 1.3 | 0.2 |
| | | | Underwear with short sleeves and legs, shirt, trousers, vest, jacket, coat, socks, shoes | 1.5 | 0.23 |

changed may be kept. It is then possible to exclude the data set from the subsequent analyses. This might apply when there had been a marked change in activity within say 15 minutes before the vote was taken. Note that the effects of large changes in activity can persist for an hour and more.

### 8.2.3 Accuracy of personal variables

In the field situation we have to rely on tables of values measured in laboratories that are listed by clothing type (clo) or activity (met) (Tables 8.1 and 8.2). This means that the values of clothing insulation and metabolic rate used are approximate and this must be remembered when interpreting the results. The accuracy of the estimate of the insulation of a clothing ensemble from tables is not high – the error is perhaps around ±20 per cent. It is a source of considerable random error in the estimation of PMV and similar indices. Estimates of metabolic rate have a similar error, but for high metabolic rates and unusual activities the error can be as much as 50 per cent. It is a source of error in predictions of comfort based on heat balance assumptions (see Chapter 4). It is more useful to make a description of the tasks being performed by your subjects, and a description of their clothing than just the estimates of met and clo.

## 8.3 Subjective measures

The design of your survey questionnaire is critical. There is a lot of expertise available from the social sciences about how to design a questionnaire and perform a survey. Even if you plan to do no more than just ask for comfort votes in a standard format it is a good idea to read a book such as Fowler (2002) about survey technique, or Oppenheim (2000) on questionnaire design, interviewing and attitude measurement. Those who are only interested in standard questions about comfort will find them in a number of papers reporting on previous surveys (e.g. McCartney and Nicol, 2002) where the questions are reported. They are also the subject of an International Standard 10551 (2001).

The subjective sensation of the warmth of the subject has traditionally been measured using a seven-point scale. The subject is asked to rate his or her feelings on a descriptive scale such as the ASHRAE or the Bedford scales shown in Table 2.1 in Chapter 2.

As you might expect when using a single number to describe so complex a thing as the thermal sensation, there is considerable controversy about such scales. In general, however, they have been well tested by the very fact of their continued use. The main difference between the two scales is the inclusion of the concept of comfort in the Bedford scale. Most researchers agree, however, that subjects use the two scales in much the same way. It is also generally agreed that there is no significant improvement in accuracy to be obtained by adding more points to the scale. Miller (1956) pointed out that many such psychophysical scales seem to contain seven points and he concludes that seven probably represents the optimum number descriptions of sensation that we can distinguish.

One of the more telling criticisms of the comfort vote is that by using a descriptive scale we run the danger of overlapping with cultural use of the words. Humphreys (2008) has warned that even with considerable care being taken, there is often a difference in the use of the scales by people with different languages or cultures. Thus a person living in a cold climate might see 'warm' as having a positive connotation ('nice and warm'), while the

**TABLE 8.2** A standard list of the metabolic rates (from ISO 8996 2004, reproduced with permission)

| Class | Average metabolic rate (with range in brackets) | | Examples |
|---|---|---|---|
| | $W.m^{-2}$ | $W$ | |
| 0 Resting | 65 (55 to 70) | 115 (100 to 125) | Resting, sitting at ease |
| 1 Low metabolic rate | 100 (70 to 130) | 180 (125 to 2325) | Light manual work (writing, typing, drawing, sewing, book-keeping); hand and arm work (small bench tools, inspection, assembly or sorting of light materials); arm and leg work (driving vehicle in normal conditions, operating foot switch or pedal. Standing drilling (small parts); milling machine (small parts); coil winding; small armature winding; machining with low power tools; casual walking (speed up to 2.5 kmh$^{-1}$). |
| 2 Moderate metabolic rate | 165 (200 to 260) | 295 (235 to 360) | Sustained hand and arm work (hammering in nails, filing); arm and leg work (off-road operation of lorries, tractors or construciton equipment); arm and trunk work (work with pneumatic hammer, tractor assembly, plastering, intermittent handling of moderately heavy material, weeding, hoeing, picking fruits or vegatables, pushing or pulling lightweight carts or wheelbarrows, walking at a speed of 2.5 kmh$^{-1}$ to 5.5 kmh$^{-1}$, forging. |
| 3 High metabolic rate | 230 (200 to 260) | 415 (360 to 465) | Intense arm and trunk work; carrying heavy material; shovelling; sledgehammer work; sawing; planning or chiselling hard wood; hand mowing; digging; walking at a speed of 5.5 kmh$^{-1}$ to 7 kmh$^{-1}$. Pushing or pulling heavily loaded hand carts or wheelbarrows; chipping castings; concrete block laying. |
| 4 Very high metabolic rate | 290 (>260) | 520 (>465) | Very intense activity at fast to maximum pace; working with an axe; intense shovelling or digging; climbing stairs, ramp or ladder; walking quickly with small steps; running; walking at a speed great than 7 kmh$^{-1}$. |

inhabitant of a hot climate would say the same of 'cool' ('nice and cool'). This will tend to skew the use of the scale by confusion between comfort and hotness. To get round this effect it is advisable to add a preference vote (PREF) to the comfort vote.,C. Andamon *et al.* (2006) among others have found that in the warm climates of Southeast Asia the preferred conditions are related to 'cool' on the ASHRAE scale, rather than neutral.

The preference vote most commonly used is the three-point one suggested by McIntyre:

I would like to be

> −1: Warmer          0: No Change          +1: Cooler

Using this scale McIntyre and Gonzalez (1976) demonstrated a distinct difference between the preferences of British subjects in winter and American subjects in the summer – in the direction expected.

Another approach, used by Griffiths (1990) among others, is to use the ASHRAE scale as its own preference scale, by adding the extra question 'how would you prefer to feel?' This locates the preferred sensation at the time, and makes the use of a separate preference scale unnecessary. Using this method Humphreys and Hancock (2007) found that people often were where they would like to be on the scale, even if this differed from 'neutral'. So you cannot assume that people always like to feel 'neutral'.

One way to get round the problem of descriptive scales is to use a semantic differential scale of the type:

> Too Cold :     :     :     :     :     :     :     : Too Hot

The subject can then 'vote' by placing a mark on the line at the appropriate point. The proportionate distance along the line will give the vote.

This type of scale would be particularly useful where subjects use a variety of languages. McIntyre (1980) warns, however, that correlations using semantic differential scales are not as high as they are with descriptive scales. This is possibly because different people interpret the desired or comfortable range differently, some thinking it includes only the central box while others think that the middle five represent comfort.

Some researchers use a continuous form of the seven-point scale. A line is labelled with the seven categories, but the subject can mark anywhere along the line, whether at a category label or between labels. However, it is found that the majority of votes are placed at the labelled points.

It is important that when you introduce subjective scales to your subjects you explain that you are asking about their *feelings*. People can sometimes use the scale to give their impression of how hot 'it' (i.e. the environment) is rather than how they feel. This is not generally what you are interested in. The preference scale should be introduced in a similar manner. The question heading the subjective scale (e.g. How are you feeling just now?) is an integral part of the scale. Changing its form or wording can alter the response to it.

Your questionnaire can also ask about such things as clothing, activity and use of windows, fans and so on. An example of a brief questionnaire for repeated use is given in an appendix to this chapter. We have collected a number of such scales, not as a guide to questionnaire design but to give you some ideas about what can be done.

The collection of comfort, preference and other votes can be done automatically, possibly using the datalogger being used for the physical environment. A number of such automatic set-ups have been used (Humphreys and Nicol, 1970; Williamson and Coldicutt, 1990; Nicol and McCartney, 2001) and they have the advantage for longitudinal surveys of allowing the subjects to be prompted, and to cast their votes without the need for an interviewer. Other methods are to issue the subject with a pad of questionnaires (see example in the appendix to this chapter), or to ask for the response during an interview. It has been increasingly common for questionnaires to be presented electronically using facilities such as SurveyMonkey that allow researchers to design their own questionnaire and have it delivered to a survey population automatically.

These methods allow more complex surveys to be undertaken, with more detailed answers on things such as activity and clothing. Such complex surveys have the drawback of being more demanding and obtrusive. We recommend that for longitudinal surveys (see Chapter 9) in particular, you keep your questionnaire as simple as possible, and use automatic logging of subjective votes where feasible. Remember also that the final database of results must coordinate the subjective results with simultaneous measurements of the physical environment.

In addition to the instantaneous comfort votes it can be well worth considering asking for a more generalised, open-ended opinion from your subjects of their satisfaction with their thermal environment. This can be done before and/or after the survey, and if done after the survey period it could be specifically aimed at getting a sort of integrated evaluation of the period of the survey. In this way you can find the overall retrospective effect of the period about which you have detailed information. Not surprisingly your subjects will often have clear and perceptive ideas about what is wrong (or right) about their environment and what can be done to improve it. As well as giving you a lot of information this can help convince your subjects that you are interested in their opinions.

## 8.4 Other subjective measures

Remembering that there are other things than warmth that affect our comfort, a number of researchers have asked for responses about both the other aspects of the thermal environment such as humidity or air movement, and also the wider environment such as the visual and acoustic conditions. Such scales, of course, require that appropriate measurements such as illuminance and noise level be made. Scales similar to that shown in Table 2.1 can be devised for these other aspects, but many of them are not symmetrical, as heat/cold, and different forms of scale have to be devised. Overall comfort can also be of interest as can perceived productivity. An advantage of this more comprehensive kind of survey is that it enables an assessment to be made of the interactions between different aspects of the environment (ASHRAE, 2011).

## 8.5 Thermal behaviour

We introduced the fundamental form of the adaptive approach to comfort in Chapter 3 and in particular the role that people play in making sure they do not become uncomfortable. They avoid discomfort chiefly by their use of the controls available to them – windows, shading, fans and so on (see Section 3.6). The use of these simple controls has been included in a number of surveys (Nicol and McCartney, 2001; Haldi and Robinson, 2009; Yun and Steemers, 2008) and they have given insights into the adaptive process (Rijal *et al.*, 2011). The results enable building thermal simulations to model more realistically people's behaviour in buildings. Much of the information has been collected from questionnaires that asked simply whether, for instance, the window was open or closed and the fan on or off. From this simple information one can develop a probability distribution of how the temperature affects window-opening or the operation of a fan (Rijal *et al.*, 2011; Haldi and Robinson, 2010).

With the right instrumentation more detailed information can be collected that includes not only whether the window was open but also how wide. Such studies have great potential for developing an adaptive model. Such a model will include consideration of the constraints on the use of the control as well as the motivation for using it (Rijal *et al.*, 2011). This is a growing area of interest to researchers, modellers and designers, and will doubtless develop rapidly.

## 8.6 The comfort questionnaire

Once you have decided on what to ask and what to measure, you have to decide how to collect the data. This will require you to produce a questionnaire so that you can ask the questions you need to ask and record the answers. Designing a useful questionnaire is difficult and demanding, and entails far more than jotting down a few questions. We recommend you read a book on questionnaire design (Oppenheim, 2000). If a pilot study is not possible because of time or cost, it is always worth trying out your questionnaire on colleagues or friends to make sure that other people's interpretations of your questions are the same as your own – often they are not!

If you are asking the questions yourself it may be that you can upload them direct to a spreadsheet or paper form, but always remember that you need to collate your physical measurements and your subjective responses, so an accurate time check is important on both. Because there is a wide range of methods now available to deliver your questions (electronically through the internet, sending by post, asking the questions directly, etc), it is difficult to give detailed advice.

An example of a longitudinal questionnaire designed to be used four times a day can be found in Table 8.3 (see appendix to this chapter).

## References

Andamon, M.M., Williamson, T.J. and Soebarto, V.I. (2006) Perceptions and expectations of thermal comfort in the Philippines, *Proceedings of the Conference on Comfort and Energy Use in Buildings: Getting them right, Windsor, UK, 27–30 April 2006.*

ASHRAE (2011) Guideline 10-2011 *Interactions Affecting the Achievement of Acceptable Indoor Environments.* Atlanta, Georgia: American Society of Heating, Refrigeration and Air Conditioning Engineers.

Fanger, P.O. (1970) *Thermal Comfort: Analysis and applications in environmental engineering.* Copenhagen: Danish Technical Press.

Fanger, P.O., Melikov, A.K., Hanzawa, H. and Ring, J. (1988) Air turbulence and the sensation of draught. *Energy & Buildings* 12, 21–39.

Fowler, F. (2002) *Survey Research Methods* (3rd edn). *Applied Social Science Research Methods, vol. 1.* London: Sage.

Griffiths, I. (1990) Thermal comfort studies in buildings with passive solar features: Field studies. *Report to the Commission of the European Community, ENS35 090 UK.*

Haldi, F. and Robinson, D. (2009) Interactions with window openings by office occupants. *Building and Environment* 44(12), 2378–2395.

Haldi, F. and Robinson, D. (2010) On the unification of thermal perception and adaptive actions. *Proceedings of Conference on Adapting to Change: New thinking on comfort, Cumberland Lodge, Windsor, UK, 9–11 April 2010.* London: Network for Comfort and Energy Use in Buildings. Available at http://nceub.org.uk.

Humphreys, M. (1974) Environmental temperature and thermal comfort. *Building Services Engineer* 42, 77–81.

Humphreys, M. (1977) The optimum diameter for a globe thermometer for use indoors. *Annals of Occupational Hygiene* 20(2), 135–140.

Humphreys, M. (2008) 'Why did the piggy bark?' Some effects of language and context on the interpretation of words used in scales of warmth and thermal preference. *Proceedings of Conference on Air Conditioning and the Low Carbon Cooling Challenge, Cumberland Lodge, Windsor, UK, 27–29 July 2008.* London: Network for Comfort and Energy Use in Buildings. Available at http://nceub.org.uk.

Humphreys, M. and Nicol, F. (1970) An investigation into the thermal comfort of office workers. *JIHVE* 38, 181–189.

Humphreys, M.A. and Hancock, M. (2007) Do people like to feel 'Neutral'? Exploring the variation of the desired sensation on the ASHRAE scale. *Energy and Buildings* 39(7), 867–874.

ISO 7726 (2001) *Thermal environments: Instruments and methods for measuring physical quantities.* Geneva: International Standards Organisation.

ISO 7730 (2005) *Ergonomics of the thermal environment: Analytical determination and interpretation of thermal comfort using calculation of the PMV and PPD indices and local thermal comfort criteria.* Geneva: International Standards Organisation.

ISO 8996 (2004) *Ergonomics of the thermal environment: Determination of metabolic rate.* Geneva: International Standards Organisation.

ISO 9920 (2009) *Ergonomics of the thermal environment: Estimation of the thermal insulation and evaporative resistance of a clothing ensemble.* Geneva: International Standards Organisation.

ISO 10551 (2001) *Ergonomics of the thermal environment: Assessment of the influence of the thermal environment using subjective judgment scales.* Geneva: International Standards Organisation.

McCartney, K.J. and Nicol, J.F. (2002) Developing an adaptive control algorithm for Europe: Results of the SCATs project. *Energy and Buildings* 34(6), 623–635.

McCullough, E.A., Jones, B.W. and Huck, J. (1985) A comprehensive data base for estimating clothing insulation. *ASHRAE Trans* 91(24), 29–47.

McIntyre, D.A. (1980) *Indoor Climate.* London: Applied Science Publishers.

McIntyre, D.A. and Gonzalez, R.R. (1976) Man's thermal sensitivity during temperature changes at two levels of clothing, insulation and activity. *ASHRAE Trans.* 82(2), 219–233.

Miller, G. (1956) The magic number seven, plus or minus 2. *Psychological Review* 67, 81–97.

Nicol, F. and McCartney, K. (2001) *Final Report (Public) Smart Controls and Thermal Comfort (SCATs).* Oxford: Oxford Brookes University.

Nicol, J.F. (1974) An analysis of some observations of thermal comfort in Roorkee, India and Baghdad, Iraq. *Annals of Human Biology* 1(4), 411–426.

Oppenheim, A.N. (2000) *Questionnaire Design, Interviewing and Attitude Measurement* (3rd edn). London: Continuum.

Rijal, H., Tuohy, P., Humphreys, M., Nicol, F. and Samuel, A. (2011) An algorithm to represent occupant use of windows and fans including situation specific motivations and constraints. *Building Simulation Journal* 4(2), in press.

Rohles, F.H. and Nevins, R.G. (1971) The nature of thermal comfort for sedentary man. *ASHRAE Transactions* 77(1), 239–246.

SurveyMonkey™ Official website. SurveyMonkey.com. Available at www.surveymonkey.com/UK.

Williamson, T.J. and Coldicutt, S. (1991) Aspects of thermal preferences in housing in a hot humid climate with particular reference to Darwin, Australia. *Int J Biometeorol* 34, 251–258.

Yun, G.Y. and Steemers, K. (2008) Time-dependent occupant behaviour models of window control in summer. *Building and Environment* 43(9), 1471–1482.

Zangheri, P. (2010) Assessing thermal comfort in practice: Long-term metering and occupant surveys in four case studies. *Next Building – advances in next generation building technologies and design*, SolExpo Fair, Verona, Italy, 5 May 2010.

## Appendix to Chapter 8: An example of a longitudinal questionnaire

**FIGURE 8.6** Sample longitudinal questionnaire

| *School of Architecture*<br>*Oxford Brookes University*<br>*Oxford OX3 0BP* | *Thermal Comfort Survey*<br>*Daily checklist*<br>*British Airways – Waterside* |
|---|---|

| Your name: | Code: | Today's date: Monday, 26th June, 2000 |
|---|---|---|

| | Time | ..... am | ..... am | ..... pm | ..... pm |
|---|---|---|---|---|---|
| **FEELINGS**    **at present I feel:** | | | | | |
| Much too cool | | | | | |
| Too cool | | | | | |
| Comfortably cool | | | | | |
| Comfortable | | | | | |
| Comfortably warm | | | | | |
| Too warm | | | | | |
| Much too warm | | | | | |
| **PREFERENCE**    **I would prefer to be:** | | | | | |
| Much cooler | | | | | |
| A bit cooler | | | | | |
| No change | | | | | |
| A bit warmer | | | | | |
| Much warmer | | | | | |
| **CLOTHING**    **(tick as appropriate)** | | | | | |
| Short sleeve shirt/blouse | | | | | |
| Long sleeve shirt/blouse | | | | | |
| Vest | | | | | |
| Trousers/long skirt | | | | | |
| Shorts/short skirt | | | | | |
| Dress | | | | | |
| Pullover | | | | | |
| Jacket | | | | | |
| Long socks | | | | | |
| Short socks | | | | | |
| Tights | | | | | |

| | | | School of Architecture<br>Oxford Brookes University<br>Oxford OX3 0BP | | Thermal Comfort Survey<br>Daily checklist<br>British Airways – Waterside | | | |

| | Time | ..... am | ..... am | ..... pm | ..... pm |
|---|---|---|---|---|---|
| **CLOTHING (continued)** | | | | | |
| Tie | | | | | |
| Boots | | | | | |
| Shoes | | | | | |
| Sandals | | | | | |
| Other (specify)............................ | | | | | |

| | Time | ..... am | ..... am | ..... pm | ..... pm |
|---|---|---|---|---|---|
| **ACTIVITY**   **in the last 15 minutes** | | | | | |
| Sitting (passive work) | | | | | |
| Sitting (active work) | | | | | |
| Standing relaxed | | | | | |
| Standing working | | | | | |
| Walking indoors | | | | | |
| Walking outdoors | | | | | |
| Other (specify)............................ | | | | | |

| | Time | ..... am | ..... am | ..... pm | ..... pm |
|---|---|---|---|---|---|
| **CONTROLS**   **tick as appropriate** | | | | | |
| Door open | | | | | |
| Windows open | | | | | |
| Blind/curtains down | | | | | |
| Lights on | | | | | |
| Air condition on | | | | | |
| Heating on | | | | | |
| Fan on | | | | | |
| Heater on | | | | | |
| Other (specify)............................ | | | | | |

# 9

# CONDUCTING A FIELD SURVEY

This chapter will give you some tips and examples to help you to design and conduct your field survey.

## 9.1 Choosing a subject population and their environment

Let's assume that you have a general idea of the group of people you want to investigate. The idea may be centred on the particular climate or country you are interested in, a type of building you want to investigate, or someone asking for help to improve their building; or it may be simply that an opportunity for a survey presents itself. You or your collaborators will then need to identify a particular group of individuals who are willing to provide the subjective responses essential for the survey – for instance the workers in an office block, the inhabitants of a group of houses or the students in a university hall of residence. The building and the room or rooms your subjects inhabit are almost as important to the survey as are the human subjects themselves. We are, after all, interested in the relationship between the two. If the room is atypical, or the building of unusual design, it will limit the applicability of your results, but it may be just what you need to answer the research question you have set yourself.

### 9.1.1 Treat your subjects right: confidentiality, ethics

Whatever method is used for data collection, the subjects will need to be briefed on the aims and methods of the survey. They will, after all, be contributing a considerable amount of effort to your project, and they need to be clear what is expected of them and what pitfalls to avoid – such as rushing in from another room with a different environment just before recording their comfort vote.

You should always make it clear that their answers are confidential and will not be passed on to anyone else except in an anonymous form. In any survey using human subjects you have to obtain the consent of your local ethics committee. This has become an increasingly important consideration and there have been instances where journals have refused to publish papers because ethics approval had not been obtained. Certainly, if you work in a university or government department, ethics clearance will be demanded. If you are from any other background – say an architectural practice – you should make sure that your subjects are not asked to give information they are unhappy about giving, and that any results you obtain are kept in a secure place.

## 9.2 Choosing a subject sample

You will need to select a sample (which may include the whole population of your building, but will normally be a subset) and make sure the individuals in the sample are willing to devote the effort needed for taking part in the survey. Any employer involved will need to agree to the workers taking part and, where appropriate, their Trade Union should also be contacted.

In choosing a sample of subjects, it is important to remember that the people chosen should be familiar with their surroundings and the climate they are living in, and that the experiment should enable them to go about their usual routine and to dress to suit themselves.

There are two basic forms of survey sampling that have been used. These are known as transverse and longitudinal sampling. In the transverse survey the whole or a substantial proportion of a population each give a single comfort assessment. In the longitudinal survey each of a small number of subjects gives a large number of comfort assessments over an extended period of perhaps several days or weeks.

### 9.2.1 Transverse sampling

The merit of the transverse survey is that:

1   It uses the whole or a large part of the chosen population and sampling bias is reduced or avoided.
2   Enough subjects are involved to ensure the results will be representative.

This method also minimises the disruption to the lives and work of the subjects. It has been used in a large number of surveys (e.g. Bedford, 1936; Schiller, 1990).

The problem with this method is that if the survey is completed in a short time, say a day, then the variety of environmental conditions you investigate with any one group of subjects is generally small, since the variation of temperature in a single day is limited. Even if there is some variation, you will be getting the reaction of a different subject to each value of the temperature.

Alternatively, if you are looking at a number of buildings, or rooms in a building, then the surveys may take some days (or even weeks) to complete. The adaptive model tells us that the subjects' responses will have changed over this period of time. Because conditions vary between one day and the next it could be difficult to distinguish between the effect of changed conditions and of the way different sets of individuals respond to them. This limits the usefulness of the transverse survey for research into adaptive comfort.

### 9.2.2 Longitudinal sampling

The longitudinal sample has been used in a number of surveys (e.g. Webb, 1964; Humphreys and Nicol, 1970; Nicol, 1974; Sharma and Ali, 1986) and has the merit that it provides a considerable body of information about a particular person or group of people, and can follow their responses over an extended period. Although many subjects can be used, considerations such as the number of volunteers you can muster or the number of instruments you can afford will generally mean that your sample is small.

This form of experiment means we can look at the difference between individuals, and with an appropriately designed experiment, the effect of serial changes in temperature on comfort. One drawback of the longitudinal survey is that it requires dedication on the part of the subjects, particularly if, as is desirable, the survey is extended beyond working hours. Most surveys in the past have been conducted only during working hours (exceptions are Webb, 1959; Nicol, 1974; Sharma and Ali, 1986; Nicol et al., 1994) and this has two drawbacks: first, the range of environments the subjects experience during waking hours may not be fully represented, particularly in climates with big diurnal temperature variations; second, time series are very hard to follow if only part of the day is covered.

Another problem with the longitudinal survey, if the number of subjects is small, is the danger of sampling bias. The sample may not be typical of the whole of the chosen population. This problem is reinforced by the dedication needed to take part, because people who volunteer are self-selected. Nevertheless it is important, as far as possible, to make your sample representative of the chosen population in such things as sex, age and weight. In this way a relatively small group can be representative of a much larger population.

> *Humphreys and Nicol (1970) carried out a survey over a period of 15 months based on hourly records collected from a small group of subjects, the records later being divided into monthly batches. They found that subjects differed from themselves from month to month by about as much as they did from one another. Such information could not have been obtained from the results of a transverse survey. The same experimental design was later used by Fishman and Pimbert (1982).*

### 9.2.3 Repeated transverse surveys

An approach that combines the virtues of longitudinal and transverse surveys is to use what might be called the 'repeated transverse survey', where the same population is visited repeatedly, say once a month or once a season for a year. This will ensure a range of different environments as the year progresses, particularly in climates with distinct seasons.

This method was used in the Pakistan Surveys reported by Nicol et al. (1999) and in the European SCATs project (Nicol and McCartney, 2001) where monthly surveys were undertaken in offices using as far as possible the same subjects each time. During the SCATs project a chosen subsample took part in longitudinal surveys, each subject providing responses up to four times a day for as long as they were willing to do so.

### 9.2.4 Other approaches

We have based our discussion on the classic survey, in which subjects are asked for their comfort vote at intervals and the environmental parameters are measured at the same time. Other experimental designs may be used.

> *Humphreys (1972) conducted a series of experiments using time-lapse photography to observe a class of school students. He used as the measured response the proportion of students who were wearing*

*minimum clothing (shirt sleeves) as against those wearing a pullover or jacket. This enabled him to get clothing 'responses' from a whole population on multiple occasions without disrupting their lives. While he was unable to follow any particular student, he was able to draw conclusions about the time-series effect of temperature-variation on clothing. Applying adaptive theory also enabled comfort information to be extracted. Modern digital and video technology make this sort of approach increasingly easy to use but again the ethics issues are different today and must be taken into account whilst developing the survey methodology.*

## 9.3 How many observations from each subject and how many subjects?

It is not possible to be precise about the number of observations needed from each subject or the number of subjects needed. In general one can say that for statistical analysis the more data, the better, but against this must be set the amount of work needed to complete the records, and the problems that arise from continuing the experiment for too long. A survey with say 20 subjects each giving 100 data sets could provide a useful amount of information. If you can get more data sets without wearing out your subjects, or unduly extending the period over which the data is collected, then do so. The larger the population the more precise your answer will be. Figure 9.1 (Humphreys *et al.*, 2010) shows how the precision of an estimate of comfort temperature from a survey varies with the number of observations in the data set, in the context in which the data were collected. About two thirds of computed values of the comfort temperature will be within ± one standard error (s.e.) of the estimated value, and 95 per cent within ±2 standard errors. So from Figure 9.1, if the survey data set has 20 comfort votes the s.e. is 0.4K, and so the error in the estimate is almost certainly less than 1K. For 50 votes the s.e. is 0.25K, and the result is probably within half a degree of the real value. So it is better to get 30 data sets than 15, and better five subjects than one. Beware, however, that there is a danger of 'subject fatigue' where they become increasingly likely to react in an automatic and less considered way to the survey questions.

For a transverse survey the number of subjects will be the same as the number of observations, so that 100 subjects might be the aim though fewer may suffice. The number of subjects you need depends on the accuracy you aim to achieve.

Some observers have made very substantial collections of data:

> *In the SCATs surveys the aim was to get about 25–30 subjects each month in each of five offices in each of the five countries taking part. In some cases the only offices available had only five workers, but with repeated surveys over a period of a year a useful quantity of data was collected from all buildings. A total of 4,655 data sets were collected. A subset of the subjects took part in longitudinal surveys, collecting between them more than 25,000 data sets, using the brief questionnaire shown as Appendix 1 in Chapter 8 (Nicol and McCartney, 2001).*
>
> *Sharma and Ali (1986), over three hot seasons in Northern India, took an average of 280 data sets for each of 18 subjects, giving a total of 5,100 data sets. Given that the subjects were working in eight hour shifts throughout the 24 hours and that they were measuring the environment with traditional instruments, the survey represents a quite prodigious effort.*
>
> *Humphreys and Nicol (1970), using a central datalogger with automated (punch tape) output, collected 5,349 data sets from 18 subjects during working hours, the data sets being divided into monthly records from each subject, with an average of 63 sets per record.*

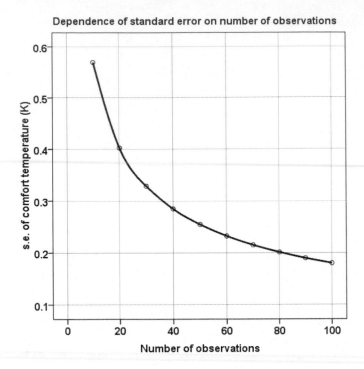

FIGURE 9.1 Standard error of the estimate of comfort temperature for a batch of n observations. The vertical axis is the standard error of the estimate of the comfort temperature ($T_c$) (K). The horizontal axis is the number of observations in the survey

Source: Humphreys *et al.*, 2010.

> *Busch (1990) obtained 1,146 data sets, all from different subjects and in four different buildings. In his 1992 paper he analyses the differences between the four different sets of subjects.*
>
> *Schiller (1990) obtained 2,342 sets of observations in ten buildings in two seasons. Although there was some overlap of subjects this was not used in the analysis. This survey suffered from one problem mentioned above: the conditions were very uniform not only within buildings but between buildings and between seasons.*
>
> *Williamson and Coldicutt (1991) collected 7,447 data sets between 78 regular respondents.*

But don't assume that large sample populations are always necessary to get useful results:

> *Griffiths (1990) conducted transverse surveys in 17 buildings or building types with passive solar features. The sample sizes in each building type were only between 14 and 53. Yet he was able to get very useful and statistically significant evidence that empirical and calculated values for the comfort temperature were different.*

## 9.4 Time sampling

It is necessary to consider time sampling as well as subject sampling because time plays such an important part in the adaptive process. By time sampling we mean the times of day and year at which a survey is carried out. The relationship between us and our thermal environment is essentially dynamic. We respond not only to the conditions at the specific moment, but also to our experience of conditions over the time leading up to it (Nicol, 1992). This means that the response of the subject to the environment in the present is influenced by their experience of previous thermal environments.

There is evidence (Nicol, 1992) that people adapt almost completely in a week or two to a change of temperature. So your subjects will be adapting themselves as the survey progresses. Consider what this means for the comfort temperature. If you are trying to find the comfort temperature in a particular set of conditions, keep your survey as short as possible, consistent with the collection of enough data.

This problem is different if it is the change of comfort temperature with time that interests you. If possible the survey should take place throughout the day and evening and not just the working day so that time series in the subjective response can be investigated. It probably also will give you a greater range of indoor conditions and make analysis easier. Such experiments can extend to periods of several days, weeks or months.

## 9.5 The data set

The concept of the data set comes from the traditional comfort survey, and it is useful to look at it. The data set is a set of readings: comfort vote, temperatures, humidity and air velocity. The values for a particular data set are taken simultaneously – or as near simultaneously as instrumentation allows. (The comfort vote should be taken first, so that any knowledge of the temperature, or the change in temperature since the last set, does not influence the subject's comfort vote.)

The frequency with which data sets can be taken is governed by the likely rate of change in thermal sensation. There is a danger of serial correlation (i.e. where a response depends on the previous response) if people are again asked to answer the same question too soon. If we regard the human body as a meter, then the rate at which that meter can change its reading is fairly slow. Just as it would be pointless, for instance, to take readings every minute from a globe thermometer that takes ten minutes to settle, so it would be pointless to take 'readings' (i.e. comfort votes) from a person at intervals of less than half an hour (unless the purpose of the experiment was to quantify the human response time). Too small an interval between the votes and we are in effect measuring the same thing again. Votes taken at too long an interval would mean we got less information than we could have obtained. An acceptable interval between subjective responses in a longitudinal survey is one hour, but in a survey in a work situation, asking subjects to respond more than four times a day is unreasonable, except possibly with respondents who are rewarded somehow for the collection of the data.

To take readings of the environmental variables only every hour, when modern dataloggers are capable of storing large amounts of data, would be to throw away information that might be useful. Maybe the subject is reacting to conditions as they were ten minutes ago, rather than at the instant of measurement. (This would be interesting to explore.) So there is a case for more frequent logging of the physical environment now that dataloggers are so versatile. It is worth bearing in mind the speed of reaction of the particular instrument. The anemometer, if it is of the fast-acting type, will need sampling more frequently than the globe thermometer. What intervals you choose will also depend on the capabilities of your logging equipment.

Some modern monitoring systems enable the collection of data streams from a number of wireless sensors that can be accessed for remote real-time display through broadband internet. The sensors could measure the physical environment, energy used and occupant use of controls such as windows, doors, lights, shading devices (blinds curtains), fans, cooling and heating. Shading use can be monitored with motion-detection cameras and the presence or absence of occupants by motion detectors. All these developments can increase the speed and accuracy of data capture, reduce errors introduced by manual transcription of data, and even ensure that any fault in the logging system can be detected before there is any serious data loss.

In addition, many surveys now use online questionnaires where the options are presented on-screen at the desk of the subject. Of course the host organisation must be willing to have the necessary software uploaded to their network. Such a set-up works well with the online environmental monitoring outlined above and will allow the collection of large numbers of responses. Online questionnaires also have the advantage of easy presentation of 'branching' questionnaires, where, for instance, the perceived reason for a particular problem could be explored. (An example might be 'do you have access to a window?' If the response is 'no', then the program passes to the next question. If it is 'yes', then the program asks 'how often do you use it?') However, there is an advantage in the researcher actually asking the subject for the responses. It ensures that the researcher is well acquainted with the subjects and the circumstances of the survey, and that they can be sure the respondent understands the question. (Of course, the question must be put to the different subjects in the same format and wording.)

In addition to monitoring the immediate environment of the subject, it is often useful to measure the conditions in the room in general. If you are going to use the results to relate comfort to overall conditions in the room, perhaps for the purpose of comparison with a computer simulation of the building's performance, then the average conditions in the room will be of interest rather than the immediate microclimate of the subject.

Outdoor conditions should also be measured, particularly if you are attempting to relate comfort to climate. These measurements are covered in Section 9.6.3.

## 9.6 Taking the measurements

### 9.6.1 The thermal environment of the subject or subjects

There is always a problem in taking measurements of a person's thermal environment, because the presence of the subject will affect the value of the immediate microclimate. Thus a globe thermometer near the subject will pick up the radiation from the subject's body as well as from the room surfaces. The air movement near the subject will be affected by the presence of the subject and so on. But if measurements are taken some way distant from the subject, how can we know that the measured conditions are those the subject experiences?

> Schiller (1990) got round this by replacing the subject by a set of instruments directly after their comfort vote had been given. The instrument set was ingeniously housed in an office chair to minimise the effect of the change in physical surroundings.

*But this raises two questions: first, the person can be replaced by a chair, but the effect of their presence – the heat plume referred to above, the breathing, etc. – cannot be replaced; second, the method is only useful for a transverse survey – a longitudinal survey where the subject is continually asked to move from their chair for the measurements to be made is not consistent with 'minimum disturbance to their normal routine'.*

Traditionally the physical measurements are taken close – but not too close – to the subject; about a metre away is usual. This method entails the observer going from subject to subject in the survey and measuring the environmental variables after having asked for a comfort vote. Generally, such surveys are restricted to the working environment.

In some surveys (Webb, 1959; Nicol, 1974; Sharma and Ali, 1986) the subjects themselves took the measurements. This is not entirely satisfactory since it risks the subjects being influenced by the readings they are taking, if the measurements are manually recorded. When using this method the subjective responses should always be completed and recorded before the environment is measured. Self-collection is unavoidable if data are to be taken in the homes of a number of subjects, unless each is allocated a technical assistant, which would be intrusive.

*The survey by Humphreys and Nicol (1970) was made in an office environment and the centrally controlled environmental monitors were placed on the desk of the subject, close to them, but not so close that the subject would unduly affect the measurements. The subject was cued to cast a comfort vote by a quiet buzzer (loud enough to attract attention but not so loud as to call the subject from another room) which was silenced by pressing one of seven buttons, one for each of the votes on the Bedford scale.*

While this was a great step forward in that the subjects were entirely unaware of the values of the temperatures and other parameters that were being measured, it still restricted the survey to the workplace, because at that time the measuring instruments had to be physically connected to the central datalogger. Modern miniaturised dataloggers capable of collecting substantial amounts of data and storing it electronically give the possibility of another step forward – they enable the logger to be with the subject to collect data 24 hours a day, wherever the subject may be (see below).

*Busch (1990) used a datalogger, well described by him as a 'toolbox' in size and general appearance. He carried the box from subject to subject, taking measurements of the environment within five minutes of the subject making a comfort assessment. The data were stored electronically. The logger was used in effect as a notebook. Schiller (1990) (see above) used a datalogger in a similar way, though the experimental set-up was rather more sophisticated. Williamson and Coldicutt (1991) used a form of datalogger to record the local environment. Here the logger was put in a permanent position in each of 31 houses in the humid tropics. The subjects were inhabitants of the houses. Up to four subjects in any one house were invited to record their comfort vote, clothing and activity up to ten times per day. The research was aimed at monitoring the houses (of which about half were air conditioned) rather than the individuals. Each individual was encouraged to use the logger at a time of day that had been identified as a problem time by that particular person (this would make analysis of differences between subjects difficult). Here again the logger is used as a notebook, so relieving the subject of the task of keeping a record of the environment and the responses to it.*

One way to conduct a survey is to use a datalogger to monitor the environment (and if possible the comfort vote and preference) of a single subject. The subject could carry the logger around for a period and at intervals the readings would be uploaded into a computer for analysis. The aim will be to monitor the personal environment of the subject and to interfere with his or her life as little as possible. Two possible methods are suggested: one uses a smaller version of Busch's toolbox which the subject carries around and places in a position close by; the other, used in the EC PASCOOL project (Baker, 1993), had the logger and instruments attached to the subject. In the longitudinal surveys carried out by Nicol *et al.* (1994) the subjects carried an automatic logger measuring temperature, air velocity and humidity. There may be difficulties with these methods: the detached monitor can be placed inappropriately or left behind; or the attached version can be too influenced by the proximity of the human body and give unreliable readings for the environment. Each hour or so, the logger will give a signal for the subject to make a response and perhaps note the clothing and activity. If possible these assessments would be entered into the logger, rather than noted on paper.

It is an informative exercise for the researcher to make such a thermal comfort record of himself or herself for a period of a week or two. It reveals the extent of the different environments encountered, the nature of one's responses to them and the practical difficulty of the task. It can be seen as a personal environmental 'calibration'.

### 9.6.2 Measuring conditions in the room

Different experiments or surveys will be concerned with different things. It may be that you are interested in the thermal performance of the building as a whole and not just a single room.

> *Roaf (1988) found that different rooms in buildings on the central Persian desert had quite different thermal characteristics, and the strategy of the occupants of the building, in moving from room to room at different times of day, was to a large extent decided by the thermal profile (the nature of the thermal environment as it changed in space and time) of each of the spaces concerned, within the thermal landscape of the building. In such a building the thermal performance of the different rooms in relation to each other as well as the outside temperature will be of particular interest.*

Another experiment might be trying to relate the temperature in the occupied space to computer predictions and relating these in turn to the thermal response of the occupants. You will need to be clear about the exact meaning of 'room temperature' used by the particular thermal simulation package you are using. This may be the operative temperature (see Section 8.1.4) or some other combination of air and radiant temperatures. The matter is discussed by Humphreys (1974) in relation to the 'environmental temperature' used for calculations using the admittance method. Also, be sure about the definitions used by the package for radiant or surface temperature: are walls assumed to be all at the same temperature? Is any particular surface assumed to be isothermal? All these factors need consideration when deciding what measurements you need to take.

The 'personality' or behaviour of an indoor climate will depend to some extent on the nature of the design of the space, the materials used and their location in the structure of the building. Thus a well-insulated closed room will tend to have equal air and radiant temperature and little air movement. Other buildings such as the Indonesian house (Figure 6.6 in Plates) rely on air movement to provide comfort. Buildings in hot dry climates such as the

Baghdadi house (Figure 6.2) provide a variety of indoor climates from which the occupant can choose as they migrate between spaces over the course of a day or year. The house's thermal mass stabilises room temperature, and in the heat of the day the walls will be cooler than the ventilation air, while at night they will be warmer, causing people in summer to sleep on the roof. We can attempt to measure the character of the room in two ways: by measuring the climate at a point in the centre of the room using a cluster of instruments and also by measuring the temperature at a number of points distributed around the room. To understand how comfort is provided in particular buildings it is important not only to understand the behaviour of the thermal landscape of the indoor climate over time, but also the thermal pathways that people take through it over time in their customary lifestyles.

### Air temperature

The air temperature in a room can vary substantially from place to place. In particular, there can be marked vertical layering of air temperature in a room. Humphreys (1972) found that the temperature measured near ceiling level in a school classroom with warm air convector heating was as much as 10K higher than the temperature experienced by the children sitting on the floor. Clearly, if we are interested in the conditions experienced by the occupants of a room, the vertical height at which the sensor is placed should be representative of the occupants' experience. It is also important to avoid placing your sensor in up- or down-draughts, which can occur near heated or cooled surfaces, unless the occupant is also subject to them. However well shielded your sensor is you should avoid places where the sun might fall directly on it. Your qualitative investigation of air patterns in the room (see Section 8.1.4) should help you to decide, but in general the air temperature should be measured not less than half a metre from any wall.

   The air temperature for a seated subject should be measured at a vertical height of about 0.6m above the floor – half the height of a seated person (different occupations may mean a different measurement height is appropriate). Often these measurements will have to be made on the subject's desk, for instance in an office setting. The temperature should preferably be sampled in other parts of the room if circumstances allow. If possible, one of these points should be near the centre of the room. The exact choice of the measurement site should be left to the judgement of the experimental team in the light of room geometry and layout. With a transverse survey the researcher will often carry the instruments and ask the questions at the workstation of the subject (e.g. Busch, 1990; Schiller, 1990; Nicol *et al.*, 1999; Nicol and McCartney, 2001; Indraganti, 2010). Here it is important that the measuring instrument is given time to settle at the measurement site before readings are taken. The settling time will depend on the instruments being used and it would be worthwhile to do a small experiment to see how long your particular set of instruments takes to settle.

### Surface temperatures

The problems of measuring the radiant temperature have been discussed already. Unless the there is a reason for wishing to know the radiant temperature, it is unnecessary to estimate it. Simply use the globe temperature to represent the room temperature for the occupant. Again it is important to use judgement in positioning the instrument, so that it is representative of the sort of place that occupants might use (and of course avoids the direct sun). If the room is large or the conditions within it vary from place to place then more than one instrument may be needed. If you are interested in the mean radiant temperature then the globe thermometer will need to be accompanied by an air thermometer and an anemometer and the measurements taken will need to be accurate.

Surface temperatures in a room can be sampled using an infra-red camera or a surface temperature 'gun'. These can be used to build up a three-dimensional picture of the radiant temperatures in the room (see Section 8.1.3). Detailed consideration of radiant asymmetry is beyond the scope of this handbook. Those interested can refer to ISO 7726 (2001) or to McIntyre (1980).

### Humidity

The water vapour pressure varies little from place to place in most rooms. A kitchen or a very damp cellar with trickle ventilation might be an exception but in most reasonably well ventilated rooms a single measurement will cover the whole space. If a number of rooms are being monitored it is worth checking a sample to see how much variation there is from room to room.

### Air velocity

As we suggested in Chapter 8, start by doing a visualisation of the normal air currents in the room using a smoke puffer or other visualisation technique. This will help you choose useful places to put an anemometer – remembering that it is the air movement that the occupants will encounter that matters. In some rooms air movement is dominated by the air entrained by moving occupants, such as when children move around in a classroom.

### 9.6.3 Measuring the conditions outdoors

We may be interested in relating preferred temperatures to the climate the subjects live in and the weather at the time of the survey. In one sense the climate they live in is defined by meteorological data collected at a local weather station. It is, after all, only these readings that we will have at our disposal when we decide what comfort temperatures to recommend. Nevertheless, it is useful to investigate the outdoor microclimate at the site of our experiment, if only to compare it to the readings at the local weather station.

> *Nicol and Kessler (1998) found that the air temperature close to the wall of a building was typically 2K higher than that measured simultaneously in a meteorological screen close by, especially when the wall was being warmed by the sun. This probably is also true of the air that enters the room through an open window.*

Measurements of outdoor air temperature, and wind speed and direction, are the most important. If possible, measurements of solar radiation onto the horizontal should be obtained. Solar intensity onto the walls of the building will clearly be useful where thermal modelling of the building is envisaged. It is possible to obtain specially designed automatic meteorological datalogging equipment to do the job. Alternatively they can be computed from the horizontal intensity using solar geometry. The exact method for taking meteorological measurements and siting the instruments is beyond the scope of this book.

A variety of measures of outdoor temperature have been used to relate it to the indoor comfort temperature. Humphreys (1978) used the monthly mean of the outdoor temperature (because this was often the only measure available to him); so does the Nicol graph (Chapter 6) for predicting suitable indoor temperatures in buildings. ASHRAE Standard 55 (2010) uses the 'prevailing mean outdoor temperature' in its 'adaptive' recommendations for temperatures in buildings (see Section 5.2.2). More recently the SCATs project (Nicol and McCartney, 2001) used an exponentially weighted running mean of the daily mean outdoor temperature. Details of the method of calculating this are found in Section 3.9.7 and in the CIBSE Guide A (CIBSE, 2006). This method is also used in the European Standard EN15251 (CEN, 2007) for predicting the comfort temperatures in free-running buildings.

## 9.7 Lack of variation in the temperature and comfort vote

One problem that some experimenters find with comfort surveys is that the temperature often varies very little in some rooms over a period of a week or a fortnight. This can be owing to 'bad luck' with the weather, or to the design of the building you are using. This is particularly common in winter-time surveys in heavyweight heated buildings – the heating system is, after all, designed to reduce temperature variations. But even in free-running buildings temperature variations can be small. Nicol recalls that he found that in the monumentally heavyweight, naturally ventilated headquarters of the Prudential Assurance in Holborn, London, temperature variations could hardly be measured – even in summer.

The problem with getting a small temperature range is that it becomes impossible to carry out the analysis of the data using the techniques that depend on temperature variation (see Chapter 10). We have only an average subjective response to an unchanging set of conditions. The experimenter is tempted to turn up the heating or throw open a window just to get some variation. But this is against the spirit of the adaptive model and can produce the effects of unexpected temperature variation for which we have criticised climate chamber experiments.

Temperature range is another reason for extending your survey to home as well as work conditions. As well as allowing you to follow the whole thermal experience of the subject, the information contained in your analysis will be improved by the increased range of conditions. Sharma and Ali (1986) did not set out to make any particular statement about the adaptive model, so there was no particular reason to follow the full daily range of their subjects' experience, but they did specifically mention the need for a wide temperature range as one reason to extend their study outside work hours.

In Section 10.3.7 we show how the information collected in circumstances of little variation can best be analysed to obtain a comfort temperature.

## References

ASHRAE (2004) Standard 55-2004. *Thermal Environmental Conditions for Human Occupancy*. Atlanta, Georgia: American Society of Heating, Refrigerating and Air Conditioning Engineers.
Baker, N. (1993) Thermal comfort evaluation for passive cooling – a PASCOOL task. Paper presented to the *Third European Conference on Architecture*, Florence, Italy, May.
Bedford, T. (1936) *The warmth factor in comfort at work*. Medical Research Council Industrial Health Research Board, Report 36. London: HMSO.
Busch, J. (1990) Thermal responses to the Thai office environment. *ASHRAE Trans* 96(1), 859–872.

CEN (2007) Standard EN15251. *Indoor Environmental Parameters for Design and Assessment of Energy Performance of Buildings: Addressing indoor air quality, thermal environment, lighting and acoustics*. Brussels: Comité Européen de Normalisation.

CIBSE (2006) Environmental criteria for design, in *CIBSE Guide A: Environmental Design*. London: Chartered Institution of Building Services Engineers, Chapter 1.

Fishman, D.S. and Pimbert, S.L. (1982) The thermal environment in offices. *Energy & Buildings* 5(2), 109–116.

Griffiths, I. (1990) Thermal comfort studies in buildings with passive solar features: Field studies. *Report to the Commission of the European Community, ENS35 090 UK*.

Humphreys, M. (1972) Clothing and thermal comfort of secondary school children in summertime. *Proceedings of the IB Commission W45 Symposium: Thermal comfort and moderate heat stress*. London: HMSO.

Humphreys, M. (1974) 'Environmental temperature' and thermal comfort. *Building Services Engineer* 42, 77–81.

Humphreys, M. (1978) Outdoor temperatures and comfort indoors. *Building Research and Practice* 6(2), 92–105).

Humphreys, M. and Nicol, J.F. (1970) An investigation into the thermal comfort of office workers. *JIHVE* 38, 181–189.

Humphreys, M.A., Rijal, H.B. and Nicol, J.F. (2010) Examining and developing the adaptive relation between climate and thermal comfort indoors. *Proceedings of Conference on Adapting to Change: New Thinking on Comfort, Cumberland Lodge, Windsor, UK, 9-11 April 2010*. London: Network for Comfort and Energy Use in Buildings. Available at http://nceub.org.uk.

Indraganti, M. (2010) Thermal comfort in naturally ventilated apartments in summer: Findings from a field study in Hyderabad, India. *Applied Energy* 87, 866–883.

ISO 7726 (2001) Standard BS ISO 7726: Thermal environments – instruments and methods for measuring physical quantities. Geneva: International Standards Organisation.

McIntyre, D.A. (1980) *Indoor Climate*. London: Applied Science Publishers.

Nicol, F. and McCartney, K. (2001) *Final Report (Public) Smart Controls and Thermal Comfort* (SCATs). Report to the European Commission of the Smart Controls and Thermal Comfort project. Oxford: Oxford Brookes University.

Nicol, F., Jamy, G.N., Sykes, O., Humphreys, M., Roaf, S. and Hancock, M. (1994) *A survey of thermal comfort in Pakistan toward new indoor temperature standards* (report). Oxford: Oxford Brookes University.

Nicol, J.F. (1974) An analysis of some observations of thermal comfort in Roorkee, India and Baghdad, Iraq. *Annals of Human Biology* 1(4), 411–426.

Nicol, J.F. (1992) Time and thermal comfort. *Proceedings of the World Renewable Energy Congress*, Reading, 13–18 September.

Nicol, J.F. and Kessler, M.R.B. (1998) Perception of comfort in relation to weather and indoor adaptive opportunities. *ASHRAE Transactions* 104(1B), 1005–1017.

Nicol, J.F., Raja, I.A., Allaudin, A. and Jamy, G.N. (1999) Climatic variations in comfort temperatures: The Pakistan projects. *Energy and Buildings* 30(3), 261–279.

Roaf, S. (1988) The Windcatchers of Yazd, PhD thesis. Oxford: Oxford Polytechnic (now Oxford Brookes University), 204.

Schiller, G. (1990) A comparison of measured and predicted comfort in office buildings. *ASHRAE Trans* 96(1), 609–622.

Sharma, M.R. and Ali, S. (1986) Tropical Summer Index – a study of thermal comfort in Indian subjects. *Building & Environment* 21(1), 11–24.

Webb, C. (1959) An analysis of some observations of thermal comfort in an equatorial climate. *BJIM* 16(3), 297–301.

Webb, C. (1964) Thermal discomfort in a tropical environment. *Nature* 202(4938), 1193–1194.

Williamson, T.J. and Coldicutt, S. (1991) Aspects of thermal preferences in housing in a hot humid climate with particular reference to Darwin, Australia. *Int J Biometeorol* 34, 251–258.

# 10

# ANALYSIS AND REPORTING OF FIELD STUDY DATA

Obviously, a field survey does not end when you have collected the data. The crux of the matter, and to many the most enjoyable part, is the analysis of the data sets of subjective votes and environmental variables. As the adaptive approach has become more widely accepted, the sophistication of the statistical methods used to analyse field study data has increased. Much of book two in this trilogy will be devoted to this topic and in this short survey we touch only on the most widely used methodologies, explaining how to avoid common pitfalls and how to get the most out of analysing your data and reporting your carefully developed conclusions.

In this chapter we will be presenting illustrative figures using data from a number of sources, but in order to give some comparability, some of them use data from the SCATs project. This was a series of comfort surveys carried out simultaneously in five European countries using identical instrumentation and questionnaires. The surveys followed the repeated transverse surveys formulation (Section 9.2.3) with monthly surveys in each of the buildings. In our figures we generally use only the data from buildings in the UK that were in free-running mode at the time of the survey. Measurements were taken on each occasion on which subjects were asked for their subjective votes. A fuller description of the project can be found in McCartney and Nicol (2002), including the wording of the questionnaire.

## 10.1 Looking at the data

Before you start using statistics to analyse your data a lot can be gained from simply looking at the data you collected. It is often useful to plot the data out in various ways. Not all will necessarily be useful or even possible with your data and this short list is just a suggestion:

- physical variables and comfort vote against time of day
- frequency plots of comfort vote and the physical variables
- physical variables against each other
- comfort or other subjective vote against physical variables, singly and in combination.

### 10.1.1 Plots against time of day

Plotting the physical variables against time of day gives you a lot of information about the conditions your subjects have experienced. Plotting the subjective votes tells you about how they have reacted to them. There may be strong daily variation in the values, particularly of temperature, and often of air velocity. This is less likely if your subjects are in a heated or cooled building than if it is free running, but still there will be a difference between conditions inside and outside working hours. You will also get some idea of how consistent your data are from subject to subject and from day to day, and how different conditions can be from one environment to another that the subjects encounter. This may help you to decide how to analyse your data.

A time plot can also suggest the motivations behind people's behaviour. Roaf (1988) plotted the temperature simultaneously in a number of rooms in houses in the Middle East against time of day. She was able to show how the movement of the occupants between the rooms was correlated with, and presumably motivated by, the variations that occurred in the environmental conditions. The occupants were chasing the most comfortable conditions by choosing between a number of alternative environments presented by the different rooms in the house (Figure 10.1).

### 10.1.2 Frequencies

Another useful graphical analysis is to plot the frequencies of different values of the variables, both for any particular subject and for the whole study. Divide the range of each variable into parts (say one degree intervals for temperature) and use a histogram to plot

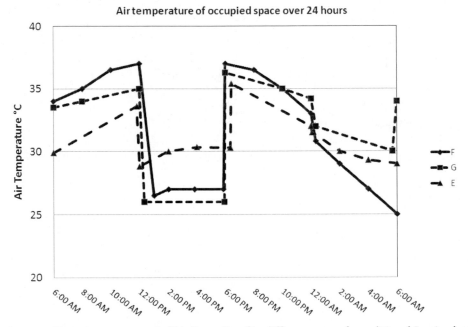

**FIGURE 10.1** Temperatures recorded by Susan Roaf in different parts of a traditional Iranian house as the occupants moved around

Source: S. Roaf (1988), redrawn by William Chan.

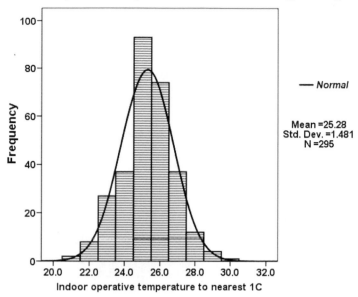

**FIGURE 10.2** Histogram of indoor operative temperatures in UK buildings in free-running mode during the SCATs project

Source: Fergus Nicol using data from SCATs project (Nicol and McCartney, 2001).

how many of your readings fall within each subdivision. Figure 10.2 shows a histogram of the frequency of temperature readings from a UK survey in free-running buildings as part of the SCATs project (Nicol and McCartney, 2001; McCartney and Nicol, 2002). The temperature was divided into groups of 1K. A normal distribution based on the mean value and standard deviation of the sample is superimposed for comparison. (The statistical programs for histograms automatically divide the variable into parts, a process known as 'grouping' or 'binning'.)

The histogram of temperatures (Figure 10.2) gives useful information about the range of conditions your subjects have experienced and the distribution of comfort votes (Figure 10.3) shows how they felt about them. Histograms also suggest the range of applicability of your results and are useful for getting a feel for the way your results are distributed. Most statistical procedures make assumptions about how the data are distributed – assuming, maybe, a normal distribution about the mean value. If your values are very asymmetrical, or even bimodal in form, then in the statistical analysis of your data you will need to bear this in mind. (A bimodal distribution has two peaks rather than just one.)

Histograms are also very useful for detecting outliers (observations that are well outside the expected range). After identifying an outlier it needs to be given careful consideration. Obvious errors can be deleted, or sometimes corrected, but an outlier may be unexpected yet correct. In this case its cause needs to be given careful consideration. It may be the origin of a fresh insight into your data.

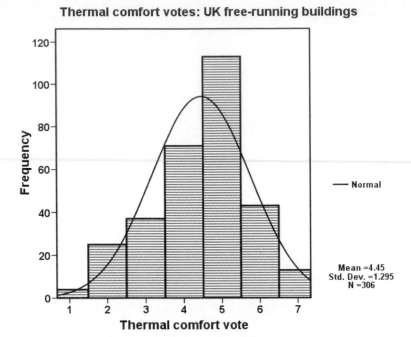

**FIGURE 10.3** Histogram of all comfort votes cast in the SCATs project in UK buildings in free-
running mode. The histogram has a superimposed normal distribution with the same
mean and assumed distribution calculated from the standard deviation

Source: Fergus Nicol using data from SCATs project (Nicol and McCartney, 2001).

### 10.1.3 Plotting the variables against each other (scatter plots)

*Operative temperature and air temperature*

The operative temperature is a combination of air and radiant temperature and can be measured approximately using a globe thermometer (see Chapters 2 and 8). As we have already
suggested, in most indoor conditions the operative temperature and the air temperature are
not very different. Figure 10.4 shows how close they are in a particular survey. Scatter plots
are also very helpful in detecting outliers.

*Air movement and air temperature*

One result of adaptive behaviour is that air velocities generally increase as the indoor temperature rises, because occupants open windows or switch on fans. Figure 10.5 shows how
air velocity changes with indoor temperature in offices in the UK, where fans are uncom-
mon, and in Pakistan (Nicol *et al.*, 1999) where they are available in practically all offices.
Interestingly, the onset of higher air velocity (caused generally by the use of fans) occurs at
temperatures of around 26°C in both countries.

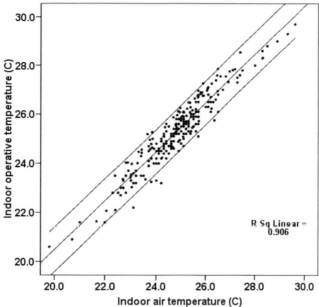

**FIGURE 10.4** Plot of mean operative temperature (measured as globe temperature with 40mm globe) against mean air temperature in UK free running buildings. Data from SCATs project (Nicol and McCartney, 2001). Mean line and 95 per cent confidence interval (within which 95 per cent of all points lie) are shown. Clearly any large deviation from the expected equality should be investigated for any anomalies in collection or measurement (for example a globe temperature in a patch of sun)

Source: Fergus Nicol using data from SCATs project (Nicol and McCartney, 2001).

## Subjective response and environment

Two obvious variables to plot against each other are comfort vote (or preference vote) and temperature (Figure 10.6). One of the most instructive things about this for those who are unfamiliar with field survey data will be how scattered the data are. Plots of comfort vote against the temperature from field surveys are characteristically dispersed in this way. This is because the comfort vote is not a precise and immediate response to temperature but is affected by all the changing environmental and social circumstances, particularly where the votes are collected over an extended period. It also differs between subjects – not everyone responds the same way. If everyone responded the same, and if comfort vote were simply a response to the environment, as it might be in a climate chamber experiment, then a closer relationship between comfort and temperature might be expected.

Also worth trying, particularly when physical theory would predict a more complex inter-relationship, is to plot comfort against two variables at the same time. There are ways in which this can be done:

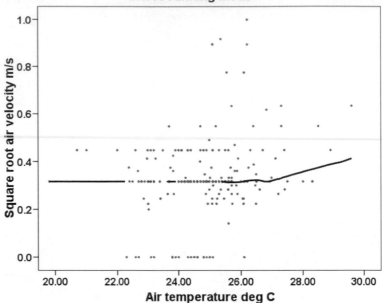

**Change of air movement with indoor air temperature: UK offices in free running mode**

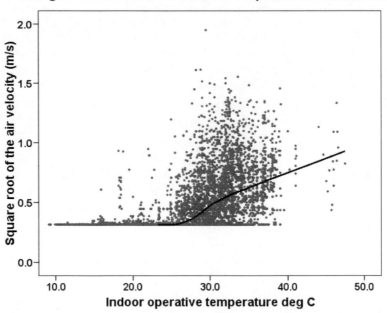

**Change of air movement with Indoor temperature: Pakistan**

**FIGURE 10.5** Variation of air velocity in a) UK and b) Pakistani offices with indoor air temperature. Data from SCATs project (Nicol and McCartney, 2001) and from data collected in Pakistani offices in 1997. The lines are Loess regression lines that show roughly the change of mean value of velocity with temperature

Source: Fergus Nicol using data from SCATs project (Nicol and McCartney, 2001).

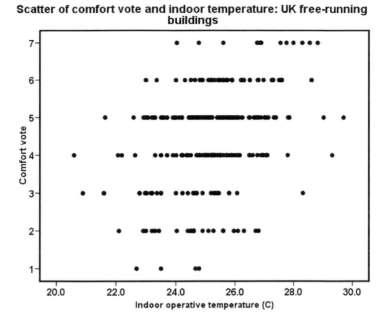

**FIGURE 10.6** Plot of comfort votes and indoor operative temperature

Source: Fergus Nicol using data from SCATs project (Nicol and McCartney, 2001).

*Nicol (1974), in his analysis of the data collected in India and Iraq, realised that air movement would have a significant bearing on the relationship between comfort and temperature. So he plotted comfort vote against temperature separately, for more than average air movement and for less than average. Placing the two scatter diagrams over one another revealed two very distinct distributions. This finding was later formalised by the use of probit analysis (see below), but the realisation that this sort of on–off distinction could be made for air movement began with a simple plot.*

Simple histograms and scatter plots can also help to guard against apparent effects that are in reality statistical fabrications. A correlation can be strongly influenced by a few outlying points, or by a particular uncharacteristic environment, or by a pair of very different environments to which the subject was exposed. Such effects will often be clear in a scatter plot in a way that could be hidden by the more sophisticated numerical outputs of a statistical program.

### 10.1.4 Proportions

A common and powerful method of analysing field study observations is to find the proportions of different comfort votes at different temperatures over the range in the survey.

*C.G. Webb (1964) devised an original way of looking at the diurnal fluctuation of comfort vote in his data from Baghdad. He realised that the day was characterised by a swing between the cool conditions in the night and the hot conditions during the day. Between these two extremes were two periods, morning and evening, when the subjects were generally comfortable. He suggested that if he identified the times of these changeover periods (Figure 10.7), and then looked at the physical conditions that characterised them, he would have a good estimate for a set of conditions that would make people comfortable. He identified the relevant times by plotting the proportion of his subjects, recording the various comfort votes against time of day. The proportion voting comfortable, comfortably warm and comfortably cool rises to a maximum at the relevant times and then falls as the proportion of subjects voting hot in the day or cool in the night increases. Figure 10.7 shows the plot he made. The maximum comfort falls at 00.00 hrs and 07.00 hrs, and he was able to read off the average values of the physical variables for these two times and make recommendations about comfortable temperatures. The results he obtained for the comfort temperature (about 34°C globe temperature) are remarkably close to those obtained by Nicol (1974) from the same data using more 'sophisticated' techniques. The method only worked for Webb because the conditions in Baghdad were consistent from day to day. A more variable climate or a less defined set of conditions would not yield a useful result. Nevertheless it demonstrates the fact that considerable information can be obtained from intelligent use of simple means and proportions.*

## 10.2 Simple statistics

Statistical packages are now widely available and spreadsheet packages will also include some statistical capabilities. This means that the actual calculation of various statistics can be done relatively easily. Below we try to look at the meaning of the numbers that the packages calculate. Those who want to learn more are referred to the simple introductions to statistics. Particularly useful are those intended for students of the social sciences, where survey work is

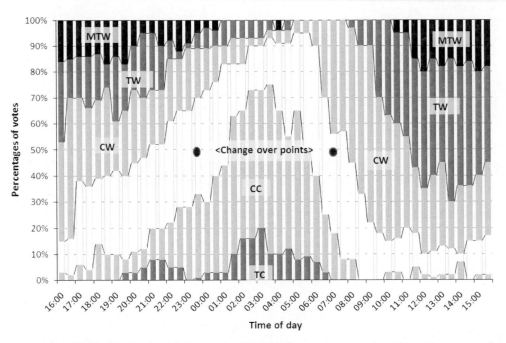

**FIGURE 10.7** Webb (1964) plotted the proportion of different votes cast by subjects in a survey in Baghdad against the time of day. Because he was collecting votes at all times of day he could identify those times when the greatest number were comfortable

Source: Fergua Nicol, from data of C.G. Webb (1964).

a principal method. Also, most statistical packages have 'help' facilities that offer explanation and tuition. If you seek help from a statistician, try to find someone familiar with the exploratory use of statistics and with the handling of field results. A book that can be recommended is Agresti and Finlay (2007). A useful book that explains statistical concepts and their application in non-technical language is Krzanowski (2007).

### 10.2.1 Means

The mean of the value of a particular variable is probably the best known statistic. In itself it should be treated with care as it can hide a lot of information about the scatter of values in the data. Scatter in your data can be tackled by calculating the means of their values over a given range. Thus one might find the mean value of comfort vote for a given range (say one degree) of temperature. This can then be plotted and a line can be drawn through it to indicate the mean response that can be expected at any given temperature. The scatter of the basic data about the line will give an indication of the variation about the mean.

### 10.2.2 Variance, standard deviations and standard errors

The sum of the squares of the difference between the mean value of a variable and its actual value divided by the number of cases minus one[1] is called the variance. The standard deviation (sd) is the square root of the variance. It is a measure of dispersion or scatter of the data. If we assume a normal distribution of the values of the variable then about two thirds of the values will lie within one standard deviation of the mean values and 95 per cent will lie within two standard deviations.

The standard error (se) of the mean is a measure of its accuracy. It is equal to the standard deviation divided by the square root of the number of observations (n). Whether or not the original data were normally distributed, the distribution of its mean tends to be normal. Thus the 95 per cent confidence limit on the value of the mean will be some two standard errors from the calculated value.

A good way of presenting these statistics for a number of subjects in a survey is to create a table of the means, standard deviations and ranges of each of the physical and subjective variables for each subject and for the group. The numbers of observations in each group should also be given. Table 10.1 is an example of such a table.

### 10.2.3 Graphical displays using means and distributions

Box plots and error bars are a simple way of displaying the changes in the mean value and distribution of one variable when another variable changes.

Box plots show the median (or middle) value of the variable as a heavy line. Around the median are shown the quartiles (so that half the values fall within the box). The extreme values are indicated by the ends of the line. All measured values of the variable except the outliers fall between the ends of the line. Finally, outliers are shown separately. Figure 10.8a shows an example where the operative temperatures shown in Figure 10.2 are presented as a series of box plots, each one representing the value of temperature measured within a particular UK building in the SCATs database.

---

1  Strictly number of degrees of freedom which in this case will be n−1 where n is the number of values.

**Table 10.1** Descriptive statistics for free-running UK buildings

| Variable | N | Minimum | Maximum | Mean | Std. deviation |
|---|---|---|---|---|---|
| Indoor operative temp °C | 295 | 20.6 | 29.7 | 25.2 | 1.5 |
| Square root of v $(m/s)^{0.5}$ | 242 | 0.00 | 1.00 | 0.33 | 0.16 |
| Relative humidity % | 265 | 30% | 87% | 44% | 7% |
| Thermal comfort votes | 306 | 1 | 7 | 4.45 | 1.295 |
| Outdoor running mean temperature ($\alpha = 0.8$) (°C) | 306 | 7.3 | 19.9 | 15.2 | 2.2 |
| PMV (calculated) | 223 | −1.05 | 1.80 | 0.44 | 0.49 |
| Comfort temperature with Griffiths constant 0.5 (°C) | 295 | 18.0 | 31.1 | 24.3 | 2.4 |
| Metabolic rate (met) | 306 | 1.00 | 2.53 | 1.34 | 0.28 |
| Clothing insulation (clo) | 306 | 0.18 | 1.07 | 0.64 | 0.15 |

The error bar shows the mean value of the variable and the range within which (say) 95 per cent of the values fall (the 95 per cent confidence interval). The error bars can show other measures of dispersion such as the standard error and the standard deviation of the values. Each error bar applies to a given subset of the data. Figure 10.8b shows an error bar plot for the same data as Figure 10.8a. Figure 10.8c shows the standard errors of the means (note the change in scale), suggesting that these distributions, though overlapping, are quite distinct.

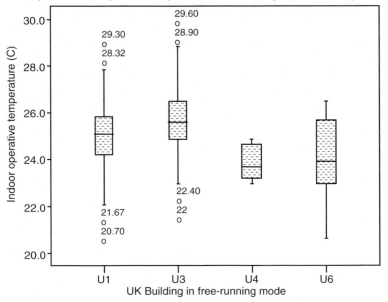

**FIGURE 10.8a** Box plots showing the mean and standard deviation of the temperature in four different UK buildings when they are in free-running mode

Source: Fergus Nicol using data from SCATs project (Nicol and McCartney, 2001).

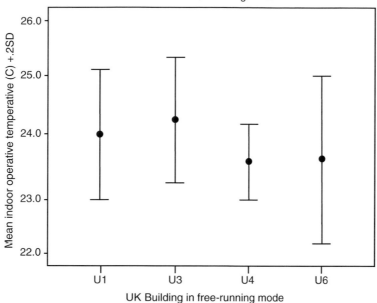

FIGURE 10.8b Error bars showing the mean and standard deviation of the temperature in four different UK buildings when they are in free-running mode

Source: Fergus Nicol using data from SCATs project (Nicol and McCartney, 2001).

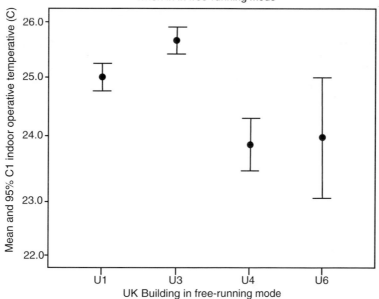

FIGURE 10.8c Error bars for mean indoor operative temperature in each UK building when in free-running mode

Source: Fergus Nicol using data from SCATs project (Nicol and McCartney, 2001).

## 10.3 More complex statistical methods

With the data from many surveys nowadays being collected automatically and often down-loaded directly into a database or spreadsheet in a computer, the researcher is able to perform statistical operations on the data relatively easily. There is therefore a temptation to rush straight into a complex statistical analysis. It is partly for this reason that we have spent some time on emphasising the advantages of taking the time to look at the data in more basic ways above. The accurate interpretation of statistical analyses requires considerable insight and experience if it is to extract reliable results from the data. A familiarity with the data will help you in this.

### 10.3.1 Correlations

The correlation between the different variables in your sample can be calculated to give a correlation coefficient that will lie between zero and one. Zero correlation means there is no relationship and the nearer the coefficient is to one the more exact is the relationship between them. The correlation coefficient indicates the strength of the relation between two of the variables – for instance temperature and comfort vote. Thus, the relative effectiveness of dif-ferent thermal indices may also be tested, the most effective index being that which is most strongly correlated with the warmth response. The Pearson's correlation, which is the one most often used, assumes that the two variables are linearly related, so two variables can be perfectly related; but if the relationship is not linear, Pearson's correlation coefficient is not an appropriate statistic for measuring their association.

A table of correlations between the variables is particularly useful in interpreting data where feedback is suspected as a factor. It is also useful to remember that the square of the correlation coefficient gives you a proportion of the variation of one variable that is explained by the other. Thus a correlation of 0.7 between temperature and comfort vote (not uncom-mon, and quite high for many surveys where temperature does not vary much) implies that 0.49 of the variation of comfort is explained by the variation of temperature. High correla-tions between the variables (for instance between $T_a$ and $T_g$) can mean that they are in effect measuring much the same thing, and to include both in a regression equation (see below) may be misleading or unnecessary.

One effect to guard against in the interpretation of correlations is data with a bipolar distribution. Such data may arise where the data come from two rather different environ-ments – say at home and at work. Since the data are in two 'bunches' it can appear to give a high correlation between comfort and temperature. If this is taken to signify a homogeneous set of data it may be misleading since in reality the subjects may react differently in the two environments as they adapt to them. Another tempting misinterpretation is to assume that a correlation between two variables implies a causal relationship. Be careful: a high correlation implies only a strong relationship and not necessarily a cause and effect. Also, if there is a cause and effect relation between two variables, the correlation cannot show which is cause and which is effect.

*Correlations between variables have been extended by Humphreys (1973) from the simultane-ous data sets we have so far been considering to time series. He was exploring the way children's clothing changed in response to classroom temperature. Instead of taking the instantaneous value of the temperature he took the running mean temperature over the previous period weighted with*

**Table 10.2** Correlations between variables and their significance in the UK free-running buildings (top right values) and for all UK buildings (lower left values) from SCATs surveys

| Correlations | C | $T_a$ | $T_g$ | RH | av | clo | met | $T_{oi}$ |
|---|---|---|---|---|---|---|---|---|
| Thermal comfort vote (C) | FR→ All↓ | 0.34 0% | 0.38 0% | −0.07 ns | −0.01 ns | −0.11 ns | 0.05 ns | 0.09 ns |
| Indoor air temperature ($T_a$) | 0.29 0.0% | | 0.95 0% | −0.52 0% | 0.19 0% | −0.20 0% | 0.02 ns | 0.17 1% |
| Indoor operative temperature ($T_g$) | 0.34 0.0% | 0.79 0.0% | | −0.44 0% | 0.17 1% | −0.18 0% | 0.06 ns | 0.18 0% |
| Relative humidity (RH) | −0.09 0.2% | −0.14 0.0% | −0.08 0.5% | | −0.18 0% | 0.04 ns | −0.05 ns | −0.22 0% |
| Indoor air velocity (av) | 0.02 ns | 0.06 6.2% | 0.04 ns | 0.03 ns | | −0.02 ns | 0.10 ns | 0.17 1% |
| Clothing insulation (clo) | −0.08 0.6% | −0.17 0.0% | −0.23 0.0% | −0.07 2.3% | −0.02 ns | | −0.20 0% | −0.24 0% |
| Metabolic rate (met) | 0.05 6.2% | −0.07 1.8% | −0.08 0.7% | 0.00 ns | −0.06 5.0% | −0.21 0.0% | | 0.09 ns |
| Outdoor temperature ($T_{oi}$) | −0.01 ns | 0.19 0% | 0.33 0% | 0.30 0% | −0.04 ns | −0.38 0.0% | −0.04 ns | |

Note: N ≈ 1200 for all (bottom left), 300 for free-running (top right).

Source: Nicol and McCartney, 2001.

*a multiplier diminishing according to the distance in time from the present – the exponentially-weighted running mean (Section 3.9.9). The result is a running mean with a characteristic half-life. Humphreys found that using such a running mean value for temperature, when given a half-life of two-and-a-half hours, increased the correlation with clothing level to 0.79 as against 0.67 using the instantaneous value of the temperature.*

## 10.3.2 Linear regression analysis

One way to see how one variable changes with another is to calculate the mean of one variable for a given range of values of the other. Figure 10.9 shows an example where the mean value of comfort vote is shown for one degree 'bins' of operative temperature, using the error bar format. The mean C for each bin of operative temperature is given by the centre points of the bands. The steady rise of the comfort vote with temperature is demonstrated.

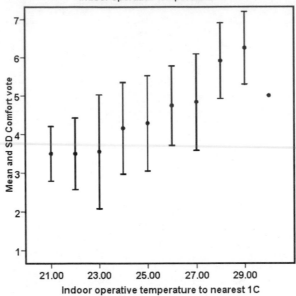

**Figure 10.9** Error bars showing the variation of mean comfort vote with one degree 'bins' of operative temperature. The circles are the mean comfort vote and the lines indicate the variation around them

Source: Fergus Nicol using data from SCATs project (Nicol and McCartney, 2001).

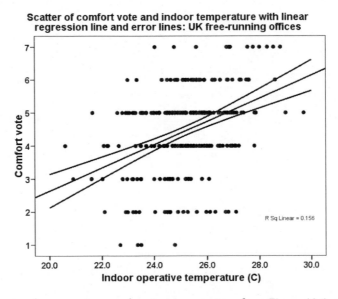

**Figure 10.10** The comfort votes measured against temperature from Figure 10.6 with regression line for C on $T_o$ added.

Source: Fergus Nicol using data from SCATs project (Nicol and McCartney, 2001).

Regression analysis is the most common method used to examine the trend of the mean response over the range of temperatures encountered. (In Microsoft Excel the regression line is called the Trendline.) Figure 10.10 shows the regression line that arises from the data shown in Figure 10.6. The standard deviation of the scatter of the votes around the regression line is called the *residual standard deviation*. The regression of comfort vote C upon operative temperature $T_{op}$ yields the line that predicts the mean response at each temperature. The extent of error in the position of the regression line is indicated by the two curves above and below the line. These lines are usually placed at the 95 per cent confidence intervals.

The relationship that links comfort and temperature can be expressed as a linear equation that predicts C from $T_{op}$. The equation of the regression line has the form

$$C = a + bT_{op} \qquad (10.1)$$

where C is the predicted comfort vote, a is the intercept on the vertical axis, b is the gradient of the line (the regression coefficient) and T is the independent variable (temperature in this case).

From the equation of the regression line the neutral temperature may be found by putting C equal to 4 (neutral or comfortable). Alternatively, it can be read from the scatter plot, from the intersection of the regression line and the C = 4 gridline. The regression line statistics are a part of the output from the SPSS statistical package. The format of the output is shown in Figure 10.11 which gives the value of a in our example as −4.037 and of b as 0.337. Note that the fact that the package gives these values to three decimal places does not guarantee that level of precision. Statistical programs usually produce the values for A and B in a table, together with their standard deviations.

There is no need to bin data for regression analysis. If data are binned, the correlation coefficient loses its meaning, and is given an artificially high value that depends on the binning interval. In the context of regression analysis, binning should be used only for illustrations, as in Figure 10.9.

Regression analysis assumes that the y-axis (comfort) is a 'dependent' variable and the x-axis (temperature) is an 'independent' variable. Thus the model assumes you are predicting comfort vote from operative temperature. Though this designation of the two variables is clear – the temperature 'causes' the subjective response – it is not altogether clear-cut in a feedback situation where the subjects will sometimes adaptively change the temperature in response to their thermal sensation.

### Coefficients[a]

| Model | Unstandardised coefficients | | Standardised coefficients | | |
|---|---|---|---|---|---|
| | B | Std error | Beta | t | Sig. |
| 1 (Constant) | −4.037 | 1.221 | | −3.307 | .001 |
| Indoor operative temperature (C) | .337 | .048 | .377 | 6.970 | .000 |

a. Dependent variable: comfort vote

**Figure 10.11** The constant term and regression coefficient for simple regression of comfort vote (C) on operative temperature ($T_o$) which is shown as a line in Figure 10.13 (output from SPSS 15.0 for windows)

Source: Fergus Nicol using data from SCATs project (Nicol and McCartney, 2001).

## 10.3.3 Multiple and curvilinear regression

Regression analysis can be extended to include more than one environmental variable by the use of multiple correlation and regression. In this way a new 'index' of thermal comfort can be constructed, as was done by Bedford (1936), Sharma and Ali (1986) and a number of other workers since. In multiple linear regression the resulting equation will have the form

$$C = a + bE_1 + cE_2 + \ldots \tag{10.2}$$

where C is the comfort vote and $E_n$ are the environmental variables.

The values of the regression coefficients a, b, c, etc. give an idea of the relative importance of the different environmental variables in deciding the value of C, while their statistical significances can indicate which variables are most important. A statistical package will generally include these in the output. The equation for a 'comfort surface' can be created by substituting the value of the 'neutral' vote for C in the regression equation.

The multiple regression equation for comfort vote of subjects in UK buildings in free-running mode is shown in Figure 10.12. The regression has three terms: operative temperature, air velocity and water vapour pressure. In this case the operative temperature is the most significant term, followed by water vapour pressure, whilst the negative term for air velocity is non-significant.

The combinations of variables on which to perform the multiple regression need to be carefully considered. If the choice is made on purely statistical grounds there is a danger that, because of the precise conditions prevailing during the survey, an equation for comfort can emerge that makes little physical sense. In this case it is unlikely that the equation will have much predictive value on any set of results other than that from which it was derived (see discussion in Sharma and Ali, 1986).

**Coefficients[a]**

| Model | Unstandardised coefficients | | Standardised coefficients | t | Sig. |
|---|---|---|---|---|---|
| | B | Std error | Beta | | |
| 1  (Constant) | −7.419 | 1.417 | | −5.237 | .000 |
| Indoor operative temperature (C) | .416 | .52 | .463 | 8.026 | .000 |
| Square root of air velocity in m/s | −.572 | .475 | −.070 | −1.204 | .230 |
| Water vapour pressure (Pa) | .118 | .045 | .147 | 2.585 | .010 |

a. Dependent variable: comfort vote

**Figure 10.12** Multiple linear regression of C on environmental variables $T_o$, $P_a$ and $V_a$. P-values suggest that temperature is the most important variable and $V_a$ the least in this set of data (output from SPSS 15.0)

Source: Fergus Nicol using data from SCATs project (Nicol and McCartney, 2001).

Regression does not necessarily need to assume a linear relationship between the variables. Thus equation 10.1 could give a quadratic relationship between comfort vote and temperature.

$$C = a + bT + cT^2 \qquad (10.3)$$

The number of regression coefficients increases and some of them may not reach significance; in the case of comfort vote and temperature a linear relationship is normally acceptable but there may be circumstances where a more complex relationship is useful.

### 10.3.3.1 A technical note on significance

Significance is often referred to in statistics and is related to what is known as the 'null hypothesis'. In this context the null hypothesis is one that says there is no relationship between the two variables. The alternative hypothesis would be that there is a relationship. The significance or p-value (sometimes just called p) is the probability of obtaining a result as extreme or more extreme than the one observed when the null hypothesis is true. So a p-value of 5 per cent would imply that, if the null hypothesis were true, the probability of observing such an extreme result would be 0.05. The probability would be higher if the alternative were true. For a given sample size the stronger the relationship between the two variables the higher the probability of a significant result. Also, the larger the sample size the more likely it is that even a weak relationship will give a significant result. Conventionally, a p-value of 5 per cent or less is taken as evidence against the null hypothesis in favour of the alternative. A p-value greater than 5 per cent only means that there is not much evidence against the null hypothesis, so it does not justify accepting the null hypothesis.

### 10.3.4 Analysis into proportions: logistic and probit analysis

As a simple example let us take a three-category comfort scale: 'too cool', 'just right' and 'too warm'. Using suitable groupings of temperature (or an appropriate thermal index) the proportions of assessments falling in each category are calculated. From these proportions a figure can be drawn (Figure 10.13). As the temperature rises the proportions finding the environment too cool will diminish and the proportion finding it too warm will increase. The proportion finding it just right will at first rise and then fall after having passed through a maximum. The temperature corresponding to the peak of the curve is usually called the 'optimum' or 'neutral' or 'preferred' temperature for the population. The curves A and B in Figure 10.13 are cumulative distributions of response to the thermal environment. They can therefore be used to estimate the variations in response among the population in the case of a transverse survey, or of individual consistency in the case of a longitudinal design. The distance AB gives for the median respondent the temperature interval corresponding to the verbal description 'just right', and if curves A and B are parallel, its centre is the neutral temperature (Humphreys, 1976).

A way to determine the statistics of the curves in Figure 10.13 is to use logistic regression. Logistic regression is particularly appropriate for analysing binary data such as we have here

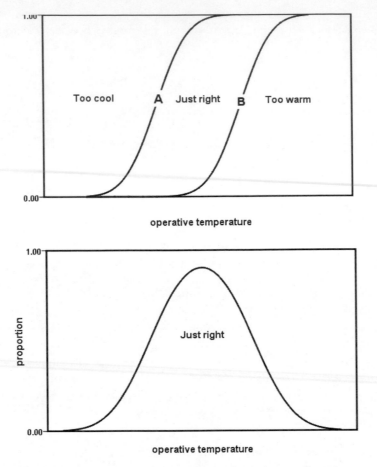

operative temperature

**Figure 10.13** Illustration of the use of probit analysis to obtain the comfort temperature

Source: Fergus Nicol.

(where say 1 = 'too warm' and 0 = 'just-right-or-too-cool'). Logistic regression predicts the probability ($p_c$) of people being 'too warm', or being 'not too cool', using the Logit function.

The Logit is defined by the relationship

$$\text{Logit } p_c = \text{Log}(p_c/1-p_c) = a + bT \tag{10.4}$$

where $p_c$ is the probability that the value of the comfort vote fulfils the description (e.g. 'the vote is "too warm"') is fulfilled, T is a measure of temperature, a is the intercept and b is the slope of the Logit line.

The value of A and B can be determined from a database of values of $p_c$ and T using linear regression and the process is known as logistic regression. Whence

$$p_c = e^{(a + bT)}/(1 + e^{(a + bT)}) \tag{10.5}$$

The resulting probability curves can be drawn using equation 10.5 once the values of A and B in the regression line have been determined.

The extension to scales having more categories is a simple matter. In comfort data there are generally seven rather than three category scales. The method will apply equally to the more complex case, except that the proportions will be, for instance, 'those voting 5 or less' rather than 'those who are not warm'. Subdivisions into the seven categories then just give us more detail about the meaning of the proportions.

Figure 10.14 shows an analysis of the seven-point comfort scale using Logit lines. Note that the lines represent, not the proportion casting a particular comfort vote but the proportion casting votes equal-to-or-less-than a particular vote. The measured proportions of subjects who actually voted in such a way are shown for comparison by the points. Figure 10.14 also shows the extension of the method to show the proportion of people comfortable or neutral. The proportion comfortable (votes 3, 4 and 5) are the proportion '5 or less' minus the proportion '2 or less'. The probability lines need not be equally spaced, since there is no reason to assume the rise in temperature associated with each rise in comfort vote has to be equal. Consideration of the meaning of an intersection, however, shows that we do have to assume the lines are parallel, and this should be taken into account when estimating the slopes of the regression lines. So all the lines will have the same slope but each will have a different intercept.

A similar but more old-fashioned approach to fitting curves such as A and B in is to use probit regression analysis (Finney, 1964). This was a method most widely used in biological experiments, particularly those concerned with establishing the effectiveness of insecticides.

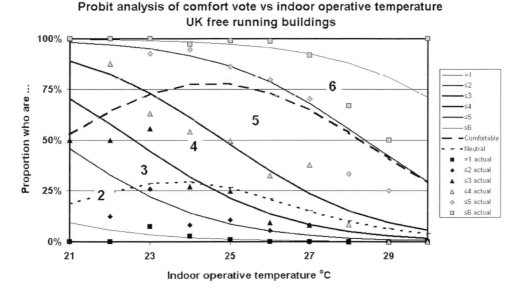

**Figure 10.14** Use of analysis to produce the proportions of subjects actually voting at different indoor operative temperatures and the curves derived from Logit regression analysis. Assuming that logit lines are parallel. Data from free-running buildings in the months April to September in UK (data from SCATs survey).

Source: Fergus Nicol using data from SCATs project (Nicol and McCartney, 2001).

**Table 10.3** Table of transformations from percentages to probits

| % | 0 | 1 | 2 | 3 | 4 | 5 | 6 | 7 | 8 | 9 |
|---|---|---|---|---|---|---|---|---|---|---|
| 0 | – | 2.67 | 2.95 | 3.12 | 3.25 | 3.36 | 3.45 | 3.52 | 3.59 | 3.66 |
| 10 | 3.72 | 3.77 | 3.82 | 3.87 | 3.92 | 3.96 | 4.01 | 4.05 | 4.08 | 4.12 |
| 20 | 4.16 | 4.19 | 4.23 | 4.26 | 4.29 | 4.33 | 4.36 | 4.39 | 4.42 | 4.45 |
| 30 | 4.48 | 4.50 | 4.53 | 4.56 | 4.59 | 4.61 | 4.64 | 4.67 | 4.69 | 4.72 |
| 40 | 4.75 | 4.77 | 4.80 | 4.82 | 4.85 | 4.87 | 4.90 | 4.92 | 4.95 | 4.97 |
| 50 | 5.00 | 5.03 | 5.05 | 5.08 | 5.10 | 5.13 | 5.15 | 5.18 | 5.20 | 5.23 |
| 60 | 5.25 | 5.28 | 5.31 | 5.33 | 5.36 | 5.39 | 5.41 | 5.44 | 5.47 | 5.50 |
| 70 | 5.52 | 5.55 | 5.58 | 5.61 | 5.64 | 5.67 | 5.71 | 5.74 | 5.77 | 5.81 |
| 80 | 5.84 | 5.88 | 5.92 | 5.95 | 5.99 | 6.04 | 6.08 | 6.13 | 6.18 | 6.23 |
| 90 | 6.28 | 6.34 | 6.41 | 6.48 | 6.55 | 6.64 | 6.75 | 6.88 | 7.05 | 7.33 |
| % | 0.0 | 0.1 | 0.2 | 0.3 | 0.4 | 0.5 | 0.6 | 0.7 | 0.8 | 0.9 |
| 99 | 7.33 | 7.37 | 7.41 | 7.46 | 7.51 | 7.58 | 7.65 | 7.75 | 7.88 | 8.09 |

Here, an underlying normal distribution, rather than a logistic distribution, is assumed for the curves A and B. The curves are converted to straight lines using the 'probit transformation'. This is done by reading off values for the proportions in each category (as a percentage) and converting them into probits using the probit table in Table 10.2 or from statistics books. Plotting the probits for comfort against the temperature (or environmental index) gives a straight-line relationship. This means a line of best fit can be estimated by eye, or a linear regression can be performed to find the positions of the probit lines. From this the category widths (such as the line AB in Figure 10.13) can be found. It is then only necessary to transform the probit values back into proportions using the table and the result is a probability curve based on statistical theory similar to that illustrated in Figure 10.14.

Both logistic and probit regression allow us to estimate the standard deviations of the intercepts and gradients of the lines (such as A and B), and therefore the confidence that can be placed in them. Webb's analysis of data from Singapore using probit analysis is a good example of the use of this method (Webb, 1959).

Probit and logistic analysis can be performed by some statistical packages without first binning the data. For logistic regression this leads to the advantage that multiple regression techniques can be used. Probit analysis is being replaced by the more convenient and versatile logistic regression methods.

### 10.3.5 Assumptions underlying simple regression and probit regression

Simple regression of comfort on temperature assumes the warmth scale to have equal intervals. This assumption is only approximately fulfilled, and certainly breaks down in the end categories of the comfort scale which have ranges that must theoretically be semi-infinite, since if a person is already 'hot' and gets hotter the scale offers no further possibilities. In most cases, however, the majority of comfort votes will be in the central categories, with only a few votes at the extremities (Figure 10.3), and the condition is often approximately met.

One advantage of probit/logistic regression analysis is that it does not require this equal-interval property. It is therefore a useful method for inspecting the intervals between the comfort descriptors on a warmth scale to see how well they fulfil the 'equal interval' assumption. An example is shown in Figure 10.14 which clearly shows that the width of the interval (as defined by AB in Figure 10.13) is wider for comfort vote 5 than it is for vote 4.

Both methods also assume freedom from auto-correlation. This means an assessment should not depend on those that precede it. In many field surveys this is unrealistic since the sensation of warmth often provokes action intended to affect the subsequent sensation. This should be remembered when the results of an analysis are interpreted.

Much the same could be said of the requirement that the properties of the population should be independent of the conditions to which they are exposed. It is in obvious contradiction to the feedback assumptions of the adaptive model. The consequence of having a population that adjusts itself in response to the comfort vote is in effect to replace the single regression line with an array of parallel lines where the higher temperatures correspond to a line shifted to the right by the lower clothing, metabolic rate, etc. which the subjects have adopted. The overall regression line will have a lower slope and a lower correlation between comfort vote and temperature. This is true of all surveys where the population is free to adapt. In extreme cases this can cause simple regression analysis to yield implausible comfort temperatures.

### 10.3.6 Errors in the regression slope or probit lines

Errors in the regression slope in linear regression, or in the intercept in probit/logistic regression will give errors in the estimates of comfort temperatures, and hence lead to error in their relationship with the outdoor temperatures. It is therefore important to understand the source of these errors. Some of the apparent errors are caused by the adaptive process itself, others by errors in the way the data are collected in field surveys and in the ways we try to relate them to environmental indices.

#### 10.3.6.1 Adaptive errors in regression analysis

As the temperature changes, people respond to the change 'adaptively' – making clothing changes, using fans or windows, etc. Over time the result of these actions is to reduce the change in comfort vote associated with a given change in temperature. Figure 10.15 demonstrates this effect when it is compared with Figure 10.10. The regression lines that apply only over a single month are steeper than those for the whole summer. Figures 10.16 and 10.11 show the same effect expressed in numbers.

#### 10.3.6.2 Adaptive errors in probit and logistical regressions

In the case of the probit or logistic regression, the effect of a population that is not homogeneous will be to flatten and depress the bell-shaped comfort curve. One group is most comfortable at one temperature and another at a different temperature, so that overall only a relatively small proportion is satisfied at both those temperatures. The range of temperatures over which at least some are comfortable widened. Similarly with a homogeneous population, but one with

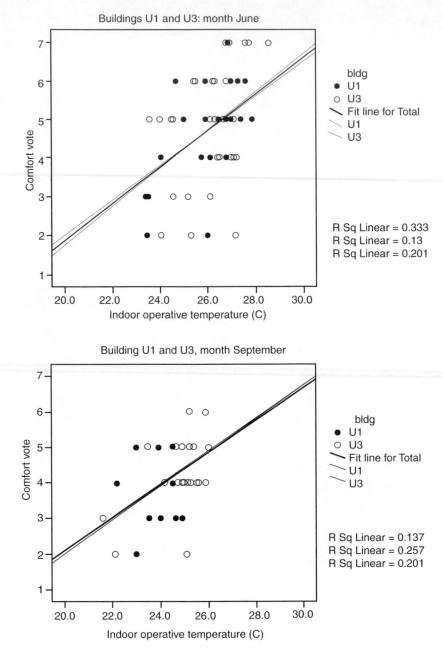

**Figure 10.15** The effect of adaptive behaviour when data are collected over an extended time. The data plotted in Figure 10.10 were collected over a number of summer months from April to September. Both figures above show only data for June and September respectively. The steeper slope for these lines are a demonstration of the effect that adaptive behaviour can have when data are collected over an extended time. The equations for lines a) and b) are shown in Figure 10.16

Source: Fergus Nicol using data from SCATs project (Nicol and McCartney, 2001).

**Coefficients[a]**

| Month | Model | Unstandardised coefficients | | Standardised coefficients | t | Sig. |
|-------|-------|------|-----------|------|------|------|
| | | B | Std error | Beta | | |
| June | (Constant) | −8.776 | 2.763 | | −3.176 | .002 |
| | Indoor operative temperature (deg C) | .522 | .106 | .515 | 4.918 | .000 |
| Sept | (Constant) | −7.586 | 2.675 | .504 | −2.836 | .006 |
| | Indoor operative temperature (deg C) | .485 | .109 | | 4.442 | .000 |

a. Dependent variable: thermal comfort

**Figure 10.16** Regression equations for the regression lines shown in Figure 10.15 (output from SPSS 15.0)

Source: Fergus Nicol using data from SCATs project (Nicol and McCartney, 2001).

thermal characteristics that vary with temperature, the curve will be 'stretched' and flattened compared with one where the population did not vary in this way.

The effect is shown by Figure 10.17. The dashed line in Figure 10.14 shows the comfort curve predicted for the whole summer. Figure 10.17 compares this with the comfort predictions using the data for June and September separately. The precision of the prediction is improved (the peak is narrower), particularly for September, and the maximum comfort level (the height of the peak) is also increased.

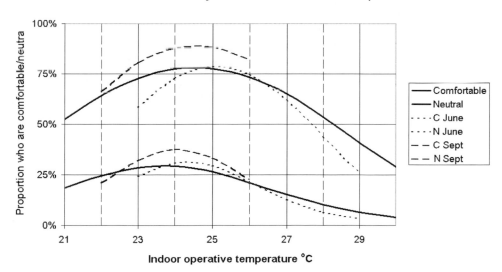

**Figure 10.17** A similar effect to that shown for regression also applies to probit analysis. The comfort predictions from the whole of the UK data are shallower and lower than for June and September alone

Source: Fergus Nicol using data from SCATs project (Nicol and McCartney, 2001).

## 10.3.6.3 Errors in the independent variable

In the classic model on which regression is based the independent variable (generally the x-axis) is assumed to be error free, all errors being associated with the dependent variable (y-axis). In real life the independent variable will contain some errors. Such errors may be of two important kinds: 'measurement errors' caused by uncertainty in the precision or accuracy of your measurements (which may be important if a number of measurements are combined); and 'equation errors' caused by the fact that the combination of factors in your multiple regression equation or thermal index such as PMV is incomplete or inappropriate. Equation error is beyond the scope of this volume, but an account of them is given by Cheng and Van Ness (1999). Measurement errors have the effect of reducing the regression slope.

These effects are small if the measurement error is small compared with the range of the variable. Measurement errors are generally small with good equipment but they can occur. Air velocity, for instance, is very variable in both time and space and precise measurements can be difficult to make since a small change in location can lead to quite large changes in measurements. If the standard deviation of the temperature is very small, the error in its measurement can appreciably reduce the slope of the regression line. When this occurs recourse may be had to the Griffiths method (see Section 10.3.7).

In our example we have used operative temperature as the predictor of thermal sensation, but there are a number of other indices that have been developed. They include Fanger's Predicted Mean Vote (PMV) (Fanger, 1970), the Corrected Effective Temperature (ET$^\star$) and Standard Effective Temperature (SET) (Gagge *et al.*, 1986). Errors in measurement or assessment of the variables from which they are calculated produce measurement errors in the index. Because these indices include a number of variables, in particular clothing insulation and metabolic rate (see Section 8.2) the cumulative error can be large. Humphreys *et al.* (2007) have gathered a number of studies where the calculated values of such indices have been correlated against the actual comfort votes of subjects. In general the more complex the index, the lower is the correlation, suggesting that increasing the supposed completeness of the index may actually introduce more error. More detail on this argument is available in Humphreys *et al.* (2007), Nicol and Humphreys (2010) give an example from the development of EN15251 of the use of various statistical techniques. Using an index that is prone to measurement error can drastically reduce the regression gradient. This leads us to look at the Griffiths method.

## 10.3.7 Griffiths' method

Griffiths (1990) introduced a neat method for assessing the mean comfort temperature for a small sample of comfort votes. It is also useful where the range of temperature is so small that regression is unreliable. Griffiths was dealing with data from buildings and groups of buildings in which he had small numbers of subjects. The data was insufficient to produce a reliable regression estimate of the comfort temperatures. Griffiths made the assumption that the increase in temperature for each scale point on the comfort scale was effectively 3K for a seven-point scale – a value well established from numerous climate chamber estimates. This means that for each comfort vote away from neutral he subtracted 3K from the actual temperature at the time to obtain the temperature that might be expected to result in neutrality. By taking the mean of these temperatures he obtained a mean comfort temperature for the sample (Figure 10.18).

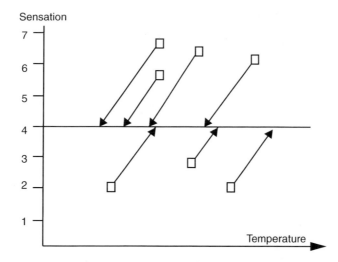

**Figure 10.18** The Griffiths method uses an assumed relationship between comfort vote and temperature so that a predicted comfort temperature (along the C=4 line) can be made for each comfort vote. The same transform can be applied to the centroid of the dataset to show the mean comfort temperature for a group of comfort votes

Source: Fergus Nicol.

A question could be raised about the use of a relation between comfort vote and temperature derived in the climate chamber. But one of Griffiths' aims was to test the reliability of climate chamber estimates of comfort temperatures in the field, and the value obtained in this way was the least controversial since it includes no adaptive effects.

### 10.3.8 Development of the Griffiths method

In effect, the method suggested by Griffiths (1990) uses a single standard value for the linear relationship between comfort vote and operative temperature: the 'Griffiths slope' G ($K^{-1}$) is equivalent to the regression coefficient. The comfort temperature can then be calculated from the comfort vote by assuming that a comfort vote of neutral will represent an estimate of 'comfort'. If the comfort vote C for neutral is $C_N$ (the numerical value of C for neutrality) then the comfort temperature $T_{comf}$ can be calculated from the actual operative temperature $T_{op}$ (or other thermal index) using the relationship

$$(C - C_N) = G(T_{op} - T_{comf}) \tag{10.6}$$

$$T_{comf} = T_{op} - (C - C_N)/G \tag{10.7}$$

Note that G can be used to estimate the comfort temperature from a single vote or from a group of votes where mean values of C and $T_{op}$ are used.

The Griffiths slope is essentially the relationship between comfort and temperature *assuming no adaptation took place*. The value of the Griffiths slope can be estimated from comfort studies and represents the maximum rate of change of comfort vote with temperature when no adaptation to temperature changes took place and measurement errors are excluded. Such a set of conditions is impossible to achieve in a real study but Humphreys *et al.* (2007) gathered together the evidence from international comfort studies (de Dear, 1998; Nicol and McCartney, 2000). Grouping them in terms of the standard deviation of the operative temperature Humphreys showed (Figure 10.19) that the regression coefficient rose to a maximum of about 0.4 when the standard deviation of temperature was about 1K. Either side of this value the regression coefficient tends to fall off. The reduction of the regression coefficient at higher standard deviation can be ascribed to the effects of adaptive actions and at lower values to measurement errors in the independent variable. Both types of error in the regression coefficient will occur even at the maximum level and so it is safe to assume that the actual value of the Griffiths slope will be greater than 0.4. Interestingly, this shows that on average people are rather more sensitive to temperature change than climate chamber studies lead us to believe.

Humphreys *et al.* (2010) further analysed the regression gradients for data from transverse surveys in which all the data are collected in a single day (a day survey), so that we can assume that little or no adaptation has taken place. Webb (1959) suggested that when performing regression analysis on the results from a number of subjects, the effects of differences between subjects could be eliminated by subtracting the mean value of each variable for that subject from the value in each data set. This method will give a more reliable estimate of the rate of change of comfort vote with changes in the environment for the 'average' individual. By adapting Webb's method to superimpose a large number of such day surveys from the

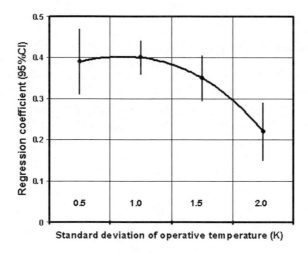

**Dependence of regression coefficient on standards deviaton of operative temperature**

**Figure 10.19** Summary of mean values of the regression coefficients (/K), in relation to the standard deviation of the operative temperature, from the de Dear (1998) and the SCAT databases. Note: The error bars indicate the 95 per cent confidence intervals of the mean of the coefficients (source: Humphreys *et al.*, 2007)

Source: Fergus Nicol.

SCATs and ASHRAE databases, Humphreys *et al.* (2010) and other surveys found that the most likely value for the Griffiths slope is about 0.5 K$^{-1}$.

### 10.3.9 Other subjective variables

The methods suggested for analysis of the variation of comfort vote according to environmental conditions can also be applied to other subjective responses such as thermal preference, and in principle to assessments of brightness, loudness of noises and intensities of odour.

## 10.4 Some common problems encountered and some mistakes to avoid

### 10.4.1 Low regression slopes

It is not uncommon to find that the regression equation for comfort vote on temperature has a low slope. As is explained above, this is natural in a system that is based on a feedback. It can, however, lead to some strange results. For instance a low slope can mean that the comfort temperature in summer, when the mean comfort vote is above neutral, appears to give a lower comfort temperature than in winter when the mean temperature is low (Figure 10.20). Data from such surveys where there was little variation in the environment and/or the comfort vote can still be used to estimate comfort temperature using an appropriately chosen

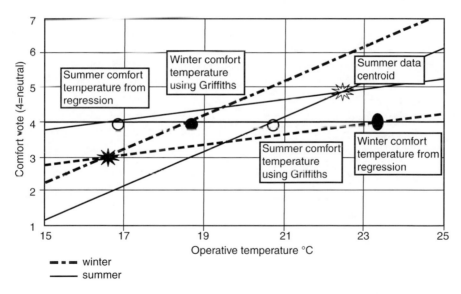

**Figure 10.20** Comfort temperatures can be misread if the regression slope is taken from a set of data with a wide spread of temperatures. This graph shows how the comfort temperature for two seasons could be miscalculated. The stars show the centroids of the data for summer and winter. If an artificially low regression slope (in this case 0.15) is calculated from the data, this means that the summer comfort temperature is lower and the winter one is higher (shown by oval markers) than they should be because the effect of adaptation is not taken into account. The values estimated using the Griffiths method are more reliable

Source: Fergus Nicol.

Griffiths regression slope for the regression line. Using the Griffiths slope of 0.5 K$^{-1}$ will give a more correct result. The Griffiths method is particularly useful if the mean comfort vote is much different from neutrality.

### 10.4.2 Confusing the dependent and independent variables

It often seems sensible to put the comfort graph the other way around with the temperature on the vertical axis and the comfort on the horizontal; in other words to regress temperature on the comfort vote. This may look sensible if you want to predict the temperature at which people return, say, a vote of comfortable. The problem is that mathematically this means that you are trying to predict the temperature from the comfort vote, which is to assert that the comfort vote is the independent variable that causes the change in temperature. There are particular circumstances where this might be defensible but for the normal prediction of comfort from temperature this will give you the wrong answer.

### 10.4.3 Taking the value of the comfort temperature from the people who vote 'neutral'

It is quite common for researchers to assume that the best temperature is the one at which people say they are comfortable. This means that they just concentrate on those data sets where the subjective vote is neutral (ASHRAE) or comfortable (Bedford). This has two drawbacks: first, you will be 'throwing out' a lot of potentially useful data; and second, because of that, you can get a wrong result. This can be seen if we consider the various values we could get for the comfort temperature from the dataset we have been using. The mean temperature at which the subjects vote neutral is 25.2°C, the regression for comfort vote on temperature predicts a comfort temperature of 23.9°C (Figure 10.10) and using the Griffiths method would give the mean comfort temperature of 24.3°C shown in Table 10.1. Thus the comfort temperature calculated from a consideration only of the temperatures at which people were comfortable gives a high estimate.

### 10.4.4 Eliminating the central category from a thermal preference scale

Some researchers, when analysing the three-category preference scale (the McIntyre scale), assign the 'no change' votes randomly to the 'prefer cooler' or the 'prefer warmer' categories before analysis. This method should be avoided for two reasons: the inclusion of a random element artificially increase the estimates of the between-subjects variability, and if the numbers of votes in the 'prefer warmer' and 'prefer cooler' categories are unequal, a biased value of the optimum temperature results.

## 10.5 Writing up your results

The way you write up your research is of course up to you, but you should remember that, as we said in the introduction to this book, thermal comfort research is a global endeavour and many of the most important findings rest on the results of a large body of studies. Your work can add to this global database of results. It is important therefore that certain basic pieces of information about your survey are given in the report, and in your published paper.

- You should always be sure to thoroughly describe your experimental aims and methodology, including the instrumentation and any precautions you take to avoid errors.
- The subjective scales you use should be reported and preferably included in the report.
- You should describe the buildings or spaces occupied by the subjects including any special features of the buildings – shading, thermal mass and so on.
- You should provide the reader with a table of results similar to Table 10.1 above. If more than one building is included in the surveys then a separate table for each building may be useful and the same applies to different seasons and especially different climates.
- Outdoor conditions should also be reported, both in terms of the overall climate and for the particular conditions during your survey.
- Always quote your results to an appropriate number of significant figures. A computer package will often produce results to several places of decimals (see for instance the regression equation in Figure 10.11). You need to think carefully what number of decimal places is justified statistically (see the standard error of the estimate in the third column) and also physically (how accurately were the variables measured).

You should remember that the reader only has your report to inform them of the context of your work as well as its results, and these need to be accurately presented whether they are your particular focus of interest or not. So, for instance, if you are interested in comparing the results from a group of vernacular buildings with those from a 'control' set of modern buildings, the reader will be interested in details of both building types.

Each individual field study is a valuable record of particular people, in particular buildings, in particular climates. Today these field studies are also particularly important because they contribute to our knowledge of trends that need to be both understood and carefully managed as we struggle to meet the multiple challenges of adapting to a changing world.

## References

Agresti, A. and Finlay, J. (2007) *Statistical Methods for the Social Sciences*. Harlow: Pearson Education.

Bedford, T. (1936) *The Warmth Factor in Comfort at Work*. Medical Research Council Industrial Health Research Board, Report 36. London: HMSO.

Cheng, C.-L. and Van Ness, J.W. (1999) *Statistical Regression with Measurement Error: Kendall's library of statistics*, vol. 6. London: Arnold.

de Dear, R.J. (1998) A global database of thermal comfort experiments. *ASHRAE Technical Data Bulletin* 14(1), 15–26.

Fanger, P.O. (1970) *Thermal Comfort*. Copenhagen: Danish Technical Press.

Finney, D.J. (1964) *Probit Analysis: A Statistical Treatment of the Sigmoid Response Curve*. Cambridge: Cambridge University Press.

Gagge, A.P., Fobolets, A.P. and Berglund, L. (1986) A standard predictive index of human response to the thermal environment. *ASHRAE Trans* 92(2), 709–731.

Griffiths, I. (1990) Thermal comfort studies in buildings with passive solar features: Field studies. Report to the Commission of the European Community, ENS35 090 UK.

Humphreys, M.A. (1973) Classroom temperature, clothing and thermal comfort: A study of secondary school children in summertime. *J. Inst. Heat. & Vent. Eng.* 41, 191–202.

Humphreys, M.A. (1976) Field studies of thermal comfort compared and applied. *Building Services Engineer* 44, 5–27.

Humphreys, M.A., Nicol, J.F. and Raja, I.A. (2007) Field studies of indoor thermal comfort and the progress of the adaptive approach. *Journal of Advances on Building Energy Research* 1, 55–88.

Humphreys, M.A., Rijal, H.B. and Nicol, J.F. (2010) Examining and developing the adaptive relation between climate and thermal comfort indoors. *Proceedings of Conference on Adapting to Change: New Thinking on Comfort, Cumberland Lodge, Windsor, UK, 9–11 April 2010*. London: Network for Comfort and Energy Use in Buildings. Available at http://nceub.org.uk.

Krzanowski, W. (2007) *Statistical Principles and Techniques in Scientific and Social Research*. Oxford: Oxford University Press.

McCartney, K.J. and Nicol, J.F. (2002) Developing an adaptive control algorithm for Europe: Results of the SCATs project. *Energy and Buildings* 34(6), 623–635.

Nicol, J.F. (1974) An analysis of some observations of thermal comfort in Roorkee, India and Baghdad, Iraq. *Annals of Human Biology* 1(4), 411–426.

Nicol, J.F. and Humphreys, M.A. (2010) Derivation of the equations for comfort in free-running buildings in CEN Standard EN15251. *Buildings and Environment* 45(1) 11–17.

Nicol, J.F. and McCartney, K. (2001) *Final Report (Public) Smart Controls and Thermal Comfort (SCATs)*. Report to the European Commission of the Smart Controls and Thermal Comfort project (Contract JOE3-CT97-0066). Oxford: Oxford Brookes University.

Nicol, J.F., Raja, I.A., Allaudin, A. and Jamy, G.N. (1999) Climatic variations in comfort temperatures: The Pakistan projects. *Energy and Buildings* 30(3), 261–279.

Roaf, S. (1988) The Windcatchers of Yazd, PhD thesis. Oxford: Oxford Polytechnic.

Sharma, M.R. and Ali, S. (1986) Tropical Summer Index: A study of thermal comfort in Indian subjects. *Building & Environment* 21(1), 11–24.

Webb, C. (1959) An analysis of some observations of thermal comfort in an equatorial climate. *BJIM* 16(3), 297–301.

Webb, C. (1964) Thermal discomfort in a tropical environment. *Nature* 202(4938), 1193–1194.

# SYMBOLS USED IN THIS BOOK

A note on temperatures: in this book temperatures in Celsius degrees are denoted by °C; temperature differences (on the centigrade scale) are denoted by K.

a,b,c...    intercept and coefficients in regression equation

Ad    Dubois surface area of the human body (m$^2$)

C    convective heat loss per square metre of body surface (W/m$^2$)

C, $C_N$    comfort vote, comfort vote for neutral

$C_{res}$    respiratory heat loss by convection (W/m$^2$)

d    diameter of a globe thermometer (m)

E    heat loss by evaporation per square metre of body surface (W/m$^2$)

$E_{is}$    insensible heat loss by evaporation (W/m$^2$)

$E_{max}$    maximum evaporation from the clothed body surface

$E_{res}$    respiratory heat loss by evaporation (W/m$^2$)

$f_{cl}$    effective clothed area (greater than one)

$f_{eff}$    effective radiation area factor (less than one)

$f_{pcl}$    permeability efficiency factor of the clothing ensemble

G    Griffiths slope (K$^{-1}$)

H    the ratio $h_c/(h_c + h_r)$ which defines operative temperature (eq. 2A.7)

$h_c$    convective transfer coefficient (W/m$^2$K)

$h_e$    evaporative heat transfer coefficient (W/m$^2$K)

$h_r$    linear radiation transfer coefficient (W/m$^2$K)

$I_{clo}$    clothing insulation (clo or m$^2$K/W) (1 clo = 0.155 m$^2$K/W)

K    conducted heat through the clothing (W/m$^2$)

L    height (m)

M    metabolic rate (met or W/m$^2$) (1 met = 57.8 W/m$^2$)

$p_c$    proportion of a population who are comfortable

$P_a$    water vapour pressure of the air (kP$_a$)

$P_{as}$    saturated water vapour pressure (kP$_a$)

$P_{ssk}$    saturated water vapour pressure at skin temperature (kP$_a$)

| | |
|---|---|
| R | radiative heat loss rate per square metre from the body surface (W/m$^2$) |
| RH | relative humidity (%) |
| S | rate of heat storage in the body (W/m$^2$) |
| $T_a$ | air temperature (°C) |
| $T_{accept}$ | range of acceptable temperatures (K) |
| $T_{cl}$ | clothing surface temperature (°C) |
| $T_{comf}$ | comfort temperature (°C) |
| $T_g$ | globe temperature (°C) |
| $T_{lim}$ | limit of acceptable temperature difference from comfort (K) |
| $T_o$ | outdoor air temperature (°C) |
| $T_{od}$ | 24-hour daily mean outdoor air temperature (°C) |
| $T_{op}$ | operative temperature (°C) |
| $T_{omin}$ | monthly mean of the minimum outdoor air temperature (°C) |
| $T_{omax}$ | monthly mean of the maximum outdoor air temperature (°C) |
| $T_{om}$ | monthly mean of the outdoor temperature (°C) |
| $T_r$ | mean radiant temperature (°C) |
| $T_{rm}$ | running mean of the outdoor daily mean outdoor air temperature (°C) |
| v | air velocity (m/s) |
| w | skin wettedness |
| W | mechanical work done (W) |
| wt | body weight (kg) |
| $\alpha$ | constant between 0 and 1 which defines the speed at which the running mean temperature responds to the outdoor temperature |
| $\varepsilon$ | emissivity of the clothed/skin surface (generally close to 1) |
| $\Theta$ | factor for balance between $T_a$ and $T_r$ in the globe temperature (equation 8.3) |

# GLOSSARY

We present here a glossary of the more common technical terms used in the study of thermal comfort. The items are brief explanations rather than rigorous definitions. Statistical terms are particularly difficult to explain clearly, briefly and accurately, and we advise the reader to refer to a good statistical textbook for full and accurate statements. We assume that the more commonly used physical terms – energy, temperature, etc., are understood. Figures in square brackets [] refer to sections in the text where you can find more. Occasionally we refer to chapters [C], tables [T] or figures [F].

**Adaptive action:** Action taken to avoid discomfort. [3.3]

**Adaptive buildings:** Buildings that give their inhabitants the opportunity to change the indoor environment to avoid discomfort. [6.3]

**Adaptive model (adaptive approach):** An approach to the study of thermal comfort that starts from the observation that there is a variety of actions people can take to achieve thermal comfort, and that discomfort is caused by constraints imposed on the range of actions by social, physical or other factors. [C3]

**Adaptive opportunity:** The opportunities provided by a building for occupants to make themselves comfortable – windows, blinds, fans, etc. [3.9.1]

**Adaptive principle:** The underlying principle of the adaptive model: *If a change occurs such as to produce discomfort, people react in ways that tend to restore their comfort.* [3.3]

**Adaptive process:** The process by which thermal comfort is improved or achieved by the use of a set of adaptive actions. [3.3]

**Air conditioning:** Though strictly defined as the control of the internal thermal environment of a building by means of controlling the characteristics of the air supplied to it by the ventilation system, AC is usually taken to mean mechanical cooling. [6.3.2]

**Anemometer:** An instrument for measuring air movement; common electrical types are the hot-wire anemometer and the hot-body anemometer. [8.1.5]

**ASHRAE database of field studies:** Database of 20,000 data sets collected from field surveys around the world. The surveys were in sufficient detail to allow for the testing of heat–balance indices using data from the field and form the basis of the adaptive standard in ASHRAE 55. [5.2.2]

**ASHRAE scale:** A set of seven (or nine) descriptors of subjective response from (Very) Hot to (Very) Cold. [T2.1]

**ASHRAE Standard 55:** A standard for the human thermal environment promulgated by the American Society of Heating, Refrigeration and Air Conditioning Engineers and based on Fanger's PMV/PPD equations. [5.2.2, F5.1]

**Bedford scale:** A set of seven descriptors of subjective response from Much Too Warm to Much Too Cool, differing from the ASHRAE scale in describing the central three categories as Comfortable. [T2.1]

**Black box:** A system or model where the inputs and the outputs are known but the internal processes which get from one to the other are not fully explained or understood. [3.6]

**Box plot:** A diagram that shows the median (middle value) of a set of values of a variable, the quartiles around it and any extreme values. [10.2.3, F10.8a]

**Building performance evaluation(BPE):** BPE involves the inspection of a building after its completion to assess the extent to which it is meeting its required performance targets and specifications. BPEs are undertaken to provide continuous feedback of evidence on performance on which strategies for performance improvements can be based (see also post occupancy evaluation). [3.2, 7.3]

**Capacitance hygrometer:** An instrument that displays the relative humidity of the air. Its active principle is the effect humidity has on the capacitance of a semi-conductor.

**Categories or classes (of buildings):** division of buildings in a standard to signify how closely they should control their indoor environment; may be a sign of supposed excellence (e.g. ISO 7730) or building type (e.g. EN15251). [5.3.1]

**CEN – Comité Européen de Normalisation:** the European standards body.

**CIBSE:** Chartered Institution of Building Services Engineers (UK).

**Climate chamber:** A laboratory in which the environmental conditions can be controlled by the experimenter; widely used to investigate the effects of the thermal environment on human subjects. [C4]

**Climate change/global warming:** changes to the global climate caused by (among other things) the pollution of the air with 'greenhouse gases' which reduce radiation heat loss from the planet to space. [C6]

**Clothing insulation:** The effective insulation of items of clothing, or a clothing ensemble, characterised as if it were a single layer covering the whole body surface; measured in clo units (1 clo = 0.155 $m^2K/W$) [2.3.6, T8.1]

**Clothing surface temperature:** Average temperature of the surface of the clothes and exposed skin; for use in heat exchange calculations. [2.3.6]

**Comfort index:** An index calculated from the attributes of the physical environment, and which attempts to predict the effect of the environment on the warmth or comfort of human occupants (see PMV, SET and empirical indices such as the Tropical Summer Index or Equivalent Temperature). [C4]

**Comfort temperature (neutral temperature):** The temperature judged by a population to be neutral on the ASHRAE scale, or Comfortable, Neither Warm nor Cool on the Bedford scale: usually assumed to be the desired temperature. [3.1]

**Comfort vote:** The subjective response given by a subject on a comfort scale such as the ASHRAE or Bedford scale. [T2.1]

**Comfort zone:** The range of temperature within which a subject will feel comfortable, though not necessarily neutral – usually taken as the three central categories on the ASHRAE or the Bedford scale. [3.4, C5]

**Compliance testing:** Testing whether the conditions provided in a building comply with those suggested by a standard. [5.2.6]

**Constraints:** Factors in the physical or social environment that hinder or prevent people from taking adaptive actions to achieve thermal comfort. [3.6]

**Convection:** The process of heat exchange between the body and the surrounding air, and taking place at the surface of the skin (or clothes) and the surfaces in the lungs. [2.3.2, F2.1]

**Coolbiz:** an initiative of the Japanese government to free up office dress codes to allow office workers to be comfortable at higher temperatures to reduce cooling energy. [F5.3]

**Core set of variables:** The minimum set of variables needed to calculate the heat exchange between the body and the environment: air temperature, radiant temperature, water vapour pressure, air velocity, clothing insulation and metabolic rate. [C8]

**Core temperature (Deep Body temperature):** The temperature of the internal organs of the body, especially the brain, the maintaining of which is the chief function of the thermoregulatory system of the body. [C2]

**Correlation:** A measure of the strength of the relationship between two variables; characterised by the **correlation coefficient**, which falls between zero and one. [10.3.1, T10.2]

**Curvilinear regression:** Regression where the relationship between the variables is not a straight line (linear regression) but a curve. [10.3.3]

**Customary environment:** An indoor environment that has become usual or normal but which will change with time (see drift of comfort conditions). [3.9, 6.1]

**Daily thermal routine:** A sequence of environments, and responses to them, which a person or group of people regularly experience during a period of 24 hours. [C1]

**Datalogger:** An instrument that accepts and stores environmental or other data electronically, ready for subsequent downloading into a computer for analysis. [C9]

**Data set:** A set of subjective and physical data obtained simultaneously. [9.5]

**Degree-days:** The product of the difference between indoor and outdoor temperature and the number of days over which it lasts, giving a rough method of estimating the likely energy use of a building. [1.2]

**Descriptive scale:** A subjective scale in which the subject is asked to choose from a given list of descriptors in casting a comfort vote. Examples are the ASHRAE scale and the Bedford scale. [8.3]

**Descriptive statistics:** a table of maximum, minimum and mean values, and standard deviations of a database of variables [T10.1]

**Dress code:** Clothing considered appropriate at the workplace, and which employees are expected to wear. [3.9.4]

**Drift of comfort conditions:** The gradual change in the comfort temperature resulting from adaptation to a changing indoor environment. [3.9.3]

**Drift of temperature:** The within-day and the day-to-day gradual changes in temperature in a room or building. [3.9]

**Dubois surface area:** An estimate of the surface area of the human body according to a formula including the person's weight and height. [2.5]

**Emissivity:** The ratio of the actual heat emitted from a surface by radiation compared with the heat emitted had it been a perfect radiator (a black body). [2.3.1, 2.5]

**Empirical model:** A comfort model derived from a statistical analysis of the data from a field survey. [C7]

**EN15251:** Adaptive standard *Indoor Environmental Input Parameters for Design and Assessment of Energy Performance of Buildings: Addressing indoor air quality, thermal environment, lighting and acoustics* produced by CEN in support of the European Performance of Buildings Directive (EPBD). [5.2.3, F5.2]

**Environmental controls:** The means by which the thermal environment can be controlled: these may require energy-use, e.g. heating or cooling systems, fans, etc, or may be 'passive', e.g. openable windows, blinds, etc. [3.6]

**Environmental variables (thermal environment):** The thermal characteristics of the environment; generally the air temperature $(T_a)$, the radiant temperature $(T_r)$, the water vapour pressure $(P_a)$ (or humidity) and the air velocity (v). [8.1]

**Equivalent temperature:** An empirical thermal index proposed by Thomas Bedford. [3.1]

**Error bar:** A line indicating the range within which a certain proportion (usually 95%) of the data might be expected to fall. [10.2.3, F10.8b, F10.8c]

**Evaporation:** The change of state from liquid to vapour. [2.3.4]

**Free-running building or free-running mode:** A building which (at the time in question) is neither being mechanically heated nor cooled. [3.5, C6]

**Fuel poverty (UK):** The state in which a person or household is likely to have difficulty in affording to keep adequately warm or cool because over 10% of disposable income is spent on fuel. [C1]

**Globe temperature:** The temperature at the centre of a black or grey-painted sphere, usually 40mm in diameter. It approximates the human body in its balance between radiant and convective heat exchange. [8.1.2]

**Greenhouse gases:** Gases present in the atmosphere that have the effect of warming the planet. There are many such gases and vapours, but carbon dioxide is of particular concern because of the inexorable rise in its concentration in the atmosphere that results from the continued burning of fossil fuels (coal, oil and gas). [C6]

**Griffiths slope:** An assumed value for the rate of change of thermal sensation with temperature, originally taken as 0.3 ASHRAE scale units per K, but other values have more recently been identified. [10.3.7]

**Heat balance model (of thermal comfort):** A thermal model of the human body built on the balance between the metabolic heat production within the body and the heat loss from the body. Heat balance is considered a necessary but not a sufficient condition for comfort. [C4]

**Heat island:** see Urban Heat Island.

**Heat loss (from the body):** The loss of heat from the human body by conduction, convection, radiation and evaporation. [2.3]

**Heat transfer coefficient:** The rate of heat flow from a surface to its surroundings, per unit area of the surface and per degree Kelvin temperature difference $(W/m^2K)$. Surface heat transfer coefficients may be for radiation or for convection. [2.5]

**Heatwave:** A prolonged period of excessively hot weather. [C6]

**Heavyweight/lightweight building:** A heavyweight building is one that has the capacity to absorb much heat in the structure. This results in steadier indoor temperatures. A lightweight building is one that can store little heat into its structure, and is therefore prone to larger swings in indoor temperature. A typical heavyweight building has masonry or concrete exposed to the interior, rather than having insulated interior surfaces. [6.1, F6.2, F6.3, F6.6, F6.7]

**HVAC (Heating Ventilating and Air Conditioning):** A term generally applied to the hardware or the industry concerned with the supply of environmental control in buildings. [5.1]

**Infra-red camera:** A camera that 'sees' surface temperatures rather than colours. It is sensitive to infra-red rather than visible electromagnetic radiation. [8.3, 9.6.1, F2.2]

**Intermediate space:** A buffer-zone between indoors and outdoors, such as a porch or a veranda. [C6]

**ISO 7730:** A temperature standard promulgated by the International Standards Organisation and based on Fanger's PMV and PPD equations. [C5]

**Kata thermometer:** A special thermometer whose purpose is to measure the cooling power of the environment, and often used as an anemometer. [8.1.5, F8.4]

**Logistic regression:** A regression model that uses the logistic distribution to predict the probability of an event occurring. The logistic distribution is similar to the normal distribution but easier to manipulate mathematically. (Contrast with least-squares regression that predicts the mean value of a variable.) [10.3.4]

**Longitudinal sampling:** A method of sampling a population in which a large number of comfort votes is sought from each of relatively few subjects (compare transverse sampling). [9.2.2]

**Mean radiant temperature:** The uniform surface temperature of a radiantly black enclosure in which a small radiantly black sphere would exchange the same amount of radiant heat as in the actual non-uniform space. (Note: definitions vary. Some refer to a person rather than a small sphere.) [8.1.3]

**Mechanical conditioning:** see air conditioning.

**Metabolic rate:** The rate of heat production within the body. The metabolic rate depends on how active a person is. It is often expressed in terms of the resting metabolism (met where 1 met = 58.2 W/m$^2$) [2.3, 8.2.2, T8.2]

**Micro-climate:** The climate in the immediate vicinity (of a building or a person). [9.6.3]

**Mixed-mode or hybrid building:** These are buildings designed to be cooled using natural ventilation from openable windows (either manually or automatically controlled), but they may also have mechanical ventilation via air distribution systems. They have refrigeration equipment if necessary for cooling during the hotter months, and heating systems if necessary for heating during the colder months. [3.7.2]

**Monitoring of buildings:** The process of recording the indoor environmental conditions and perhaps other variables such as energy consumption of a building over a substantial period of weeks or months. [9.5]

**Multiple regression:** Also called multivariate regression. A regression process that has more than one independent (predictor) variable. For example, a regression built to predict the mean comfort vote from room temperature and air velocity. [10.3.3]

**Natural ventilation:** Ventilation that relies only on air pressure differences from wind and from air temperature differences (buoyancy). (Contrast with mechanical ventilation.) [5.3.1, 6.1]

**Naturally conditioned building:** A building whose interior conditions are maintained without the use of an HVAC system. [5.3.4]

**Neutral temperature:** see comfort temperature.

**Nicol graph:** A graph that displays the seasonal changes in adaptive comfort temperatures and the seasonal changes in outdoor air temperature. It is used as a design aid. [6.3.1]

**Night ventilation, night cooling:** Night ventilation is the use of the cold night air to cool down the structure of a building so that it can absorb heat gains in the daytime thus reducing the indoor temperature during the day. [3.7.2, 6.3.1]

**Normal distribution:** A frequency distribution shows the probability that a variable will have a certain value; the normal distribution is bell-shaped, symmetrical about the mean value. It underlies much statistical analysis. For its mathematical form and properties see statistical textbooks. [10.3.3, F10.2]

**Operative temperature:** A weighted mean of air temperature and mean radiant temperature, the weights being in proportion to the convective and radiative heat transfer coefficients of the clothed human body. For practical purposes it is measured using a 40mm globe thermometer. [2.5, 8.1.4]

**Overheating:** A condition whereby a space is too hot for comfortable human occupancy. [5.6]

**Passive buildings:** Also called passive solar buildings. Passive buildings are designed to work with the climate so that the form and fabric of the building contribute significantly to the work of cooling and warming the occupants. Windows, walls and floors are designed to collect, store and distribute solar energy in the form of heat in the winter and reject solar heat in the summer with the help of locally appropriate orientation, window openings, insulation and thermal mass. This is called passive design or climatic design because, unlike active solar or mechanical heat recovery systems, it does not involve the use of mechanical and electrical devices. Not to be confused with Passivhaus design in which mechanical systems are widely used. [C6]

**Physical environment (thermal environment):** The environment in a room or building defined by its environmental variables. [2.3, 8.1]

**Post occupancy evaluation (POE):** The process of evaluating the actual and perceived performance of buildings after they have been built and occupied. The purpose is to reduce energy consumption and costs, and carbon dioxide emissions from it, while at the same time improving the comfort and acceptability of the building for its occupants (see also building performance evaluation). [3.2, 7.3]

**Predicted mean vote (PMV):** The average comfort vote predicted by a theoretical index for a group of subjects when subjected to a particular set of environmental conditions. It is primarily used in Fanger's prediction method. [4.1.1]

**Predicted percent dissatisfied (PPD):** The percentage of subject population dissatisfied (uncomfortable) in a given environment as predicted by a theoretical index. It is primarily used in Fanger's prediction method. [4.1.1]

**Preference vote:** The response of a subject to a thermal preference scale; it may be a preference for a warmer or cooler environment, or for no change. Alternatively it can be a response indicating which of the points on the ASHRAE scale would currently be the preferred state. [8.3]

**Prevailing mean outdoor temperature:** The outdoor temperature expressed as an average over a period of days or longer. A particular example is the exponentially weighted running mean temperature. [5.2.2]

**Probit analysis:** A statistical analysis whereby the proportions of people casting different comfort votes at a particular temperature can be estimated for a population. [10.3.4]

**Psychophysics:** The branch of psychology dealing with the relationship between our sensations and the stimuli from the physical world. [2.2]

**Ratchet effect:** A process in which a change is difficult or impossible to reverse (E. Shove). [6.2]

**Regression (linear):** A statistical technique by which the variation of a dependent variable can be predicted from the value of an independent variable (or variables) through a constant regression coefficient (or coefficients). [10.3.2, F10.10]

**REHVA:** The Federation of European Heating, Ventilation and Air Conditioning Associations. [5.5]

**Running mean temperature – exponentially weighted RMT:** A running mean (sometimes called a moving average) is a mean calculated for a given period of time. As time progresses, new values are incorporated in the mean, while older values are dropped from it. An exponentially weighted running mean is a special example, whereby the weight given to an item in the mean reduces according to how distant it is in the past. [3.9.9]

**SCATs database:** A database of field surveys in five European countries carried out using a standard methodology. Provides the basis for the adaptive standard in EN15251. Used for most of the illustrative material in Chapter 10. [5.2.3, F8.1]

**Scatter plot:** A graph plotting the values of one variable against the corresponding values of another variable. The 'scatter' gives a visual impression of their correlation. [10.1.3, F10.4, F10.6]

**Semantic differential:** A psychological scaling technique for obtaining a value for subjective response in which the subject is asked to denote the intensity of a stimulus by choosing a subdivision between two extremes. [8.3]

**Shoulder months:** Those months in spring and autumn where ambient temperatures tend to rise and fall relatively rapidly. [3.7]

**Significance, significance testing:** A statistical procedure to estimate the likelihood of a result occurring by chance. [10.3.3.1]

**Skin temperature:** The average temperature of the skin surface. [2.5]

**Skin wettedness:** A theoretical measure of the coverage of the skin by sweat, expressed as a proportion of the total skin surface area. It is a concept used in calculating evaporative heat exchanges, and has been linked to discomfort. [2.3.4]

**Slow buildings:** Passive buildings that rely of the use of well-designed thermal mass internally to attenuate internal swings in temperature by effectively storing excess heat in the building fabric and releasing it back into the space steadily over time. In this way the internal climate of the buildings is slowed down, peaks and troughs of temperature reduced in magnitude and higher comfort levels achieved with reduced use of heating or cooling energy. (See also: heavyweight building.) [C6]

**Standard deviation:** Expresses the variability of the set of observations. The square root of the variance of a set of data. The variance of a set of observations is the sum of the squares of the departures from their mean value divided by the number of observations less one. [10.2.2]

**Standard effective temperature (SET):** An index devised by A.P. Gagge to indicate the warmth of an environment. It is the temperature of an environment (50 per cent relative humidity, air velocity below 0.1m/s, air temperature equal to radiant temperature) in which the total heat loss from an occupant (metabolic rate 1 met, clothing 0.6 clo) would be the same as from a person in the actual environment, with their actual clothing and metabolic rate. [C4]

**Standard error:** The **standard deviation** divided by the square root of the number of observations in a set of data. It expresses the uncertainty in the estimate of the mean. [10.2.2]

**Steady-state model (of thermal comfort):** A theoretical model of thermal response based on climate chamber measurements in conditions that are held constant in time. [C4]

**Subject:** A person taking part in an experimental investigation (in our case a comfort survey). [C7, C9]

**Subject population:** A group of people who form the basis for a comfort survey. [C7]

**Subject sample:** A sample of the subject population who participate in the survey. [9.2]

**Subjective response:** The response of a subject to the experimental condition. (In our case the response is generally a comfort vote in response to a particular thermal environment.) [8.3]

**Surface temperature:** The temperature of the surface of an object or person which governs radiant heat exchange with other surfaces. [2.3.1, 8.1.3]

**Survey (comfort survey, field survey):** An experimental investigation of the subjective responses of a group of subjects in the field, generally undertaken in such a way as to disrupt the normal pattern of the subjects' lives as little as possible, and to leave subjects to decide their own dress and activity, use of environmental controls and so on. [C7, 9.2]

**Sweat rate:** The rate at which sweat is produced by the body. [2.3.4]

**Systematic error:** An error that is not mere random scatter, but indicates a bias. For example, the approximations within a theoretical comfort model can produce biases (systematic errors) in its predictions. [C10]

**Temperature standards:** Recommended values for temperature in buildings or rooms, values generally defined by the expected use of the room. [C5]

**Thermal comfort:** a) A comfortable thermal state; b) The study of processes and conditions that produce or fail to produce comfortable thermal states. [1.1.1]

**Thermal controls:** see environmental controls.

**Thermal delight:** Sensation of pleasure experienced by people who occupy optimally comfortable ambient climate conditions of temperature, humidity and air movement. [1.1.3]

**Thermal environment:** The thermal characteristics of a particular location including the radiant, convective and evaporative conditions in it. [C2]

**Thermal experience:** The different thermal environments experienced by a subject taking account of the order in which they occurred. [3.9]

**Thermal performance:** The characteristic way in which a building reacts to the thermal climate. [C6]

**Thermo-regulation:** the various physiological means by which the core temperature is regulated: vasoregulation, sweating and shivering. [2.1]

**Time sampling:** The selection of times for obtaining data. For example, sampling by time of day; a wide time sample is necessary if the full experience of subjects is to be represented. [9.4]

**Transverse sampling:** A method of choosing a population sample by taking one or a few comfort votes from each of a large number of subjects (compare longitudinal sampling). [9.2.1]

**Tropical summer index:** A thermal index derived by Sharma and Ali (1986) to express the warmth of the North Indian summer. [3.1, 9.3]

**Urban heat island (UHI):** The phenomenon that outdoor temperatures in urban areas are higher than those in the surrounding countryside, particularly at night. This is because heat absorbed from the sun and generated by people and machines is stored by buildings and other urban structures. [3.7.2]

**Vasoregulation (vasoconstriction, vasodilation):** The physiological mechanism by which the body regulates the supply of blood to the periphery of the body to reduce heat loss from the skin (vasoconstriction) or enhance it (vasodilation). [2.1]

**Warmth sensation:** Subjective response to the thermal environment. [2.2]

**Water vapour pressure:** The pressure exerted by the water vapour present in the atmosphere; **saturated WVP:** The maximum pressure that could be exerted by the water vapour present in the atmosphere at that temperature. [2.3.4, 2.5, 8.1.6]

# BIBLIOGRAPHY

Ackerman, M. (2002) *Cool Comfort: America's romance with air-conditioning*. Washington, DC: Smithsonian Institution Press.

Agresti, A. and Finlay, J. (2007) *Statistical Methods for the Social Sciences*. Harlow: Pearson Education.

ASHRAE (2009) *ASHRAE Handbook of Fundamentals*. Atlanta, Georgia: American Society of Refrigerating and Air Conditioning Engineers.

Baird, G. (2010) *Performance in Practice: Mixed mode, passive and environmentally sustainable buildings – designer and user perspectives*. London: Routledge.

CIBSE (2006) *Environmental Design: CIBSE Guide A*. London: Chartered Institution of Building Services Engineers.

Clark, R.J. and Edholm, O.G. (1985) *Man and his Thermal Environment*. London: Arnold.

Cooper, G. (1998) *Air Conditioning America: Engineers and the controlled environment 1900–1960*. Baltimore, MD: Johns Hopkins University.

Fanger, P.O. (1970) *Thermal Comfort: Analysis and applications in environmental engineering*. Copenhagen: Danish Technical Press.

Finney, D.J. (1964) *Probit Analysis*. Cambridge: Cambridge University Press.

Fisk, D.J. (1981) *Thermal Control of Buildings*. London: Applied Science Publishers.

Fowler, F. (2002) *Survey Research Methods* (3rd edn). Applied Social Science Research Methods, vol. 1. London: Sage.

Guilford, J.P. (1954) *Psychometric Methods*. New York: McGraw Hill.

Heschong, L. (1979). *Thermal Delight in Architecture*. Cambridge, MA: MIT Press.

Hopkinson, R.G. (1963) Architectural Physics: Lighting. London: HMSO.

Koenigsberger, O.H., Ingersoll, T.G., Mayhew, A. and Szokolay, S.V. (1973) *Manual of Tropical Housing and Building, Part 1: Climatic Design*. London: Longman.

Krzanowski, W. (2007) *Statistical Principles and Techniques in Scientific and Social Research*. Oxford: Oxford University Press.

McIntyre, D.A. (1980) *Indoor Climate*. London: Applied Science Publishers.

Nakicenovic, N. and Swart, R. (2010) *Special Report on Emission Scenarios*. Cambridge: Cambridge University Press.

Nicol, J.F. and Roaf, S. (2007) Progress in passive cooling: Adaptive thermal comfort and passive architecture. In M. Santamouris (ed.) *Advances in Passive Cooling*. London: Earthscan.

Oppenheim, A.N (2000) *Questionnaire Design, Interviewing and Attitude Measurement* (3rd edn). London: Continuum.

Parsons, K.C. (2003) *Human Thermal Environments* (2nd edn). Oxford: Blackwell Scientific.

Preiser, W. and Vischer, J. (2005) *Post Occupancy Evaluation*. New York: Harper Collins.

Roaf, S. (ed.) (2010) *Transforming Markets in the Built Environment: Adapting for climate change*. London: Earthscan.

Roaf, S., Crichton, D. and Nicol, F. (2009) *Adapting Buildings and Cities for Climate Change* (2nd edn). Oxford: Architectural Press.

Roaf, S., Fuentes, M. and Thomas, S. (2012) *Ecohouse: Design Guide* (4th edn). Oxford: Architectural Press.

Rudge, J. and Nicol, F. (eds) (2000) *Cutting the Cost of Cold: Affordable Warmth for Healthier Homes*. London: E&FN Spon.

Shove, E. (2003) *Comfort, Cleanliness and Convenience*. Oxford: Berg Publishers.

Stevens, S.S. (1975) *Psychophysics: Introduction to its perceptual, neural and social prospects*. New York: John Wiley & Sons.

Wright, L. (1964) *Home Fires Burning*. London: Routledge & Kegan Paul.

# INDEX